PURE AND APPLIED MATHEMATICS

A Program of Monographs, Textbooks, and Lecture Notes

Executive Editors — *Monographs, Textbooks, and Lecture Notes*

    Earl J. Taft
    Rutgers University
    New Brunswick, New Jersey

    Edwin Hewitt
    University of Washington
    Seattle, Washington

Chairman of the Editorial Board

    S. Kobayashi
    University of California, Berkeley
    Berkeley, California

Editorial Board

Masanao Aoki
University of California, Los Angeles

Glen E. Bredon
Rutgers University

Sigurdur Helgason
Massachusetts Institute of Technology

G. Leitman
University of California, Berkeley

W. S. Massey
Yale University

Irving Reiner
University of Illinois at Urbana-Champaign

Paul J. Sally, Jr.
University of Chicago

Jane Cronin Scanlon
Rutgers University

Martin Schechter
Yeshiva University

Julius L. Shaneson
Rutgers University

LECTURE NOTES
IN PURE AND APPLIED MATHEMATICS

1. *N. Jacobson,* Exceptional Lie Algebras
2. *L.-Å. Lindahl* and *F. Poulsen,* Thin Sets in Harmonic Analysis
3. *I. Satake,* Classification Theory of Semi-Simple Algebraic Groups
4. *F. Hirzebruch, W. D. Neumann,* and *S. S. Koh,* Differentiable Manifolds and Quadratic Forms
5. *I. Chavel,* Riemannian Symmetric Spaces of Rank One
6. *R. B. Burckel,* Characterization of C(X) among Its Subalgebras
7. *B. R. McDonald, A. R. Magid,* and *K. C. Smith,* Ring Theory: Proceedings of the Oklahoma Conference
8. *Yum-Tong Siu,* Techniques of Extension of Analytic Objects
9. *S. R. Caradus, W. E. Pfaffenberger,* and *Bertram Yood,* Calkin Algebras and Algebras of Operators on Banach Spaces
10. *Emilio O. Roxin, Pan-Tai Liu,* and *Robert L. Sternberg,* Differential Games and Control Theory
11. *Morris Orzech and Charles Small,* The Brauer Group of Commutative Rings

Other volumes in preparation

*The Brauer Group
of Commutative Rings*

THE BRAUER GROUP OF COMMUTATIVE RINGS

Morris Orzech   Charles Small
QUEEN'S UNIVERSITY
KINGSTON, ONTARIO

MARCEL DEKKER, INC., New York

COPYRIGHT © 1975 by MARCEL DEKKER, INC. ALL RIGHTS RESERVED.

Neither this book nor any part may be reproduced or transmitted in any form or by any means, electronic or mechanical, including photocopying, microfilming, and recording, or by any information storage and retrieval system, without permission in writing from the publisher.

MARCEL DEKKER, INC.

270 Madison Avenue, New York, New York 10016

LIBRARY OF CONGRESS CATALOG CARD NUMBER: 74-24669

ISBN: 0-8247-6261-4

Current printing (last digit):
10 9 8 7 6 5 4 3 2 1

PRINTED IN THE UNITED STATES OF AMERICA

INTRODUCTION.

The classical theory of central simple algebras over a field K concerns itself with finite dimensional K-algebras A which have K as center and admit no proper two-sided ideals. Wedderburn's description of such an algebra A, as a matrix ring $(D)_n$ with entries from a division ring D having K as center, raises the question of classifying central division algebras over K. This leads naturally to considering the equivalence relation $\sim$ on the class of central simple K-algebras, where $A \sim B$ if $A = (D)_m$ and $B = (D)_n$ for some m and n. The equivalence classes form a group with operation induced by tensor product $\otimes_K$ and inverse induced by taking the opposite algebra A^o. This group, $B(K)$, is called the Brauer group of K.

In the years surrounding 1930 a number of people investigated properties of central simple K-algebras as elements of the Brauer group. It became apparent that $B(K)$ is one of the more subtle invariants of K, and that in many cases it reflects arithmetic properties of K. An important aspect of the study of $B(K)$ is to find, for a given element $[A]$ in $B(K)$, a suitable field extension L of K (called a splitting field for A) such that $[L \otimes_K A]$ is the trivial element in $B(L)$, i.e., such that A is in the kernel $B(L/K)$ of the naturally induced map $B(K) \to B(L)$. For example, if K is an algebraic number field then each element of $B(K)$ admits a finite Galois splitting field with cyclic Galois group; this key result was proved in 1932 under the impetus of Albert, Brauer, Hasse and Noether.

Such considerations as these in the theory of the Brauer group of a field form a backdrop for much of the work generalizing that theory. We summarize the classical theory in Chapter 3.

The Brauer group $B(R)$ of a commutative ring R was introduced by Auslander and Goldman in 1960, building on work of Azumaya for R local. Their success in giving a conceptually pleasant foundation for $B(R)$ may be attributed to the happy interaction of three elements in their exposition. First, they defined the notion of an R-algebra being separable, which in the case of a finite field extension coincides with the usual notion. It turns out that an R-algebra A being central separable (i.e., A is separable and has center R)* coincides with its being central simple, when R is a field; and over any commutative ring R, suitable equivalence classes of central separable algebras can be made into a Brauer group. Secondly, building on the notion of separability they introduced, Auslander and Goldman defined Galois extensions of rings, and generalized the classical theory by giving a cohomological description $H^2(G,U(S)) \simeq B(S/R)$, for S a Galois extension of R with group G (when $Pic(S)=0$). Their work also implies that, for R local, any central separable R-algebra is split by a Galois extension. Thirdly, the Morita theory of category equivalences was brought to bear, connecting the structure theory of central separable algebras to the study of categories of modules.

In these notes we follow the approach developed in AUSLANDER & GOLDMAN [2]. Our subject is the functor which attaches to each commutative ring R its Brauer group $B(R)$. Questions about separable algebras, Galois theory of rings, etc. are pursued only insofar as they relate to the Brauer group, and the interested reader is referred to DEMEYER & INGRAHAM [1] for further details. Chapters 1, 2 and 4

---

*(in Bourbaki's terminology, A is an Azumaya R-algebra)

collect the necessary preliminaries. We create only as much machinery as we need, avoiding, for example, the full yoga of Morita theory.

Many of the basic results on separability, collected in Chapter 2, appear with simple computational proofs based on the existence of a separability idempotent e_A in $A \otimes_R A^o$ for A a separable R-algebra. The key observation is that $e_A A$ is the center of A, and the applications of this in subsequent proofs suggest themselves naturally. The advantage of avoiding the Morita theory is that the latter often tells us two objects are isomorphic, while leaving it far from clear that a certain obvious map at hand is an isomorphism. We do describe part of the Morita theory in Chapter 1 as an aid in later work, and in order to give an idea of how it plays a role in the subject. Chapters 2 and 4 contain examples and exercises which gather together material about quaternion algebras. We hope these examples will enliven the exposition, and remove from Chapter 4 the stigma of unredeemed utilitarianism.

Chapter 5 goes part of the way toward showing that over a complete local ring R the classical theory applies. We prove that if R is such a ring and m is its maximal ideal, then $B(R) \to B(R/m)$ is injective. The map is in fact an isomorphism, but we do not need the surjectivity in later chapters. As an application we prove that $B(R)$ is trivial for finite R.

Chapter 6 gives the basic theorems, due to Auslander and Goldman, about the Brauer group of a regular domain R with quotient field K : $B(R)$ is embedded in $B(K)$, and if R has dimension ≤ 2 then $B(R)$ is the intersection of the $B(R_p)$ for p ranging over the primes in R of height one. As an application we compute $B(R)$ for R the ring of integers in a number field.

It would perhaps have been worthwhile to expose the material on orders in Chapter 6 in the context of Krull domains, rather than Noetherian integrally closed domains. However, too little would have been gained in later applications in view of the digression required. Instead, we have concentrated on simplifying the traditional proofs for the material in Chapter 6, calling on as little knowledge of commutative algebra as possible. With this in mind, an appendix to Chapter 11 fills, in an elementary fashion, an important gap in Chapter 6, viz. the result of AUSLANDER & GOLDMAN [1] that if R is regular and M is a reflexive finitely generated R-module with projective endomorphism ring, then M is projective. The reader interested in orders over Krull domains is referred to BASS [2].

In Chapter 7 we introduce Galois extensions S of R and obtain the basic results about the first and second cohomology of the Galois group G. Instead of discussing the usual map $H^2(G,U(S)) \to B(S/R)$ (constructed for fields in Chapter 3) and showing it is an isomorphism when $Pic(S) = 0$, we have modified a technique of DIEUDONNÉ [2] to construct an isomorphism $B(S/R) \to H^2(G,U(S))$ under the same hypothesis. The proof that this map is a homomorphism is more conceptual than is the corresponding result for the traditional map.

Chapter 8 begins with a proof of Tsen's theorem : if F is an algebraic extension field of $K(X)$, where K is algebraically closed, then $B(F) = 0$. This, together with the Galois cohomology of Chapter 7, is used to study the map $B(R) \to B(R[X])$. Chapter 9 relates the preceding material to the question of cancelling central separable algebras : if $A \otimes X \cong A \otimes Y$ is $X \cong Y$?

Chapter 10 (an elementary discussion of faithfully flat descent) and Chapter 11 (which solves, in the traditional way, the problem of finding good splitting rings when the ground ring is local) are tools for Chapter 12. The latter contains an elementary proof, due to Knus and Ojanguren, that $B(R)$ is a torsion group for any commutative ring R. The classical proof of this theorem, when R is a field, is sketched in chapter 3, and depends on the fact that in this case $B(R)$ has a good cohomological description. In Chapter 13, the Knus-Ojanguren proof is seen to be intimately related to a cohomological description of $B(R)$ for arbitrary R.

It is a pleasure to express our thanks at this point: to Lindsay Childs, for not only noticing that Chapter 13 could be done, but also for doing it; to Gerald Garfinkel for his help with the appendix to Chapter 11 and with the material on quaternion algebras; to the students who endured various false starts and excursions in a graduate course on this material which we gave at Queen's University during 1972-73; to Karen Ede for a typing job well done; and to all the others who helped at critical points, sometimes without knowing it.

<u>Notation</u>: $\mathbb{Z}, \mathbb{Q}, \mathbb{R}, \mathbb{C}$ will have their usual meanings: the ring of integers, the rational field, the real numbers, and the complex numbers, respectively.

Morris Orzech
Charles Small

July, 1974.

CONTENTS

Introduction, iii

Chapter 1.   Preliminaries on Modules, 1
Chapter 2.   Central Separable Algebras, 11
Chapter 3.   The Brauer Group of a Field, 25
Chapter 4.   Lemmas, mostly about Separability, 37
Chapter 5.   Complete Local Rings, 49
Chapter 6.   The Brauer Group of some Special Domains, 59
Chapter 7.   Galois Cohomology, 81
Chapter 8.   Tsen's Theorem and $B(R) \to B(R[X])$, 95
Chapter 9.   Cancellation, 103
Chapter 10. Faithfully Flat Descent, 111
Chapter 11. Splitting Rings over Local Rings, 123
             Appendix, 130
Chapter 12. The Brauer Group is Torsion, 135
Chapter 13. The Full Brauer Group and Cech Cohomology
             (by Lindsay N. Childs), 147

Bibliography, 163
Notation, 179
Index, 181

The Brauer Group of Commutative Rings

CHAPTER 1.   PRELIMINARIES ON MODULES.

In this chapter we shall set out results about modules in as much generality as required for later applications. Although the presentation will be fairly self-contained, the underlying assumption is that the reader has had some prior contact with the concepts introduced. In what follows R will always be a commutative ring with 1. All rings have unit, all modules are unitary.

1.1 **Definitions.** An R-module P is called **projective** if it is a direct summand of a free R-module. Equivalently, every epimorphism $M \to P \to 0$ splits. Another characterization of a projective R-module P is that P has a projective basis $\{x_i, f_i\}$; i.e., x_i is in P, f_i in $P^* = \text{Hom}_R(P,R)$ and for each x in P, $x = \Sigma f_i(x) x_i$ (almost all $f_i(x)$ are zero). (See CARTAN & EILENBERG [1], pp. 6, 7, 132.) If P is finitely generated and projective, it admits a finite projective basis.

An R-module P is called **faithfully projective** if it is finitely generated, projective and faithful, i.e., its annihilator in R, $\text{ann}_R(P)$, is zero.

For an R-module P, $\text{tr}_R P$ is the ideal of R defined as the image of the natural map $P^* \otimes_R P \to R$ ($f \otimes x \mapsto fx$); thus $\text{tr}_R P$ is $\{fx \mid f : P \to R, x \text{ in } P\}$.

1.2 **Lemma.** Let M <u>be a finitely generated</u> R-<u>module</u>, I <u>an ideal of</u> R <u>satisfying</u> $IM = M$. <u>Then</u> $(1-a)M = 0$ <u>for some</u> a <u>in</u> I.

Proof. Let $x_1, \ldots, x_n$ be a set of R-generators of M. Then $x_i = \Sigma a_{ij} x_j$ with a_{ij} in I. Let A be the $n \times n$ matrix

$(\delta_{ij} - a_{ij})$. The classical adjoint A^* of A satisfies $A^*A = (\det A)I_n$. The equation $AX = 0$, where X is the vector $[x_1, \ldots, x_n]^t$ therefore yields that $(\det A)x_i = 0$ for $i = 1, \ldots, n$, i.e., $(\det A)M = 0$. But clearly $\det A = 1 - a$ with a in I .

1.3 **Lemma.** <u>Suppose P is faithfully projective over R . Then $\mathrm{tr}_R P = R$</u> .

<u>Proof</u>. The existence of a projective basis implies that $(\mathrm{tr}_R P)P = P$. By 1.2 , $(1-a)P = 0$, for some a in $\mathrm{tr}_R P$. Since P is faithful, $a = 1$, and $\mathrm{tr}_R P = R$.

1.4 **Corollary.** <u>Let A be an R-algebra which is a faithfully projective R-module. Then the inclusion embeds R as a direct summand of A</u> .

<u>Proof</u>. Let $g_i : A \to R$, y_i in A , satisfy $\Sigma g_i(y_i) = 1$. Then $t : A \to R$ by $t(x) = \Sigma g_i(xy_i)$ splits the inclusion.

1.5 **Proposition.** <u>Let R_1 , R_2 be commutative R-algebras. Let P_i be a finitely generated projective R_i-module, $i = 1, 2$</u> . <u>Then</u>:

(a) <u>$P_1 \otimes_R P_2$ is a finitely generated projective $R_1 \otimes_R R_2$-module. If each P_i is faithfully projective over R_i , then $P_1 \otimes_R P_2$ is faithfully projective over $R_1 \otimes_R R_2$</u> .

(b) <u>There is an isomorphism</u>

$$\mathrm{End}_{R_1}(P_1) \otimes \mathrm{End}_{R_2}(P_2) \to \mathrm{End}_{R_1 \otimes R_2}(P_1 \otimes P_2)$$

<u>given by sending $h_1 \otimes h_2$ to $h_1 \otimes h_2$ ($\otimes = \otimes_R$)</u> .

Proof. (a) If $\{x_i, f_i\}$, $\{y_j, g_j\}$ are projective bases of P_1 and P_2 over R_1 and R_2 respectively, then $\{x_i \otimes y_j, f_i \otimes g_j\}$ is a projective basis of $P_1 \otimes P_2$ over $R_1 \otimes R_2$.

Suppose P_1 and P_2 are faithful over R_1 and R_2. Then $tr_{R_i}(P_i) = R_i$, i.e., there exist $\mu_1, \ldots, \mu_n$ in P_1, $h_1, \ldots, h_n$ in $Hom_{R_1}(P_1, R_1)$ with $\Sigma h_i(\mu_i) = 1$, and similarly for P_2. It follows that $tr_{R_1 \otimes R_2}(P_1 \otimes P_2) = R_1 \otimes R_2$. It is then clear that $P_1 \otimes P_2$ is faithful over $R_1 \otimes R_2$.

(b) We shall write down the inverse of the natural map given, leaving the necessary but unpleasant computations to the reader. A conceptual proof is to be found in BASS [1], Corollary 2.11, p. 94. Let $\{x_i, f_i\}$, $\{y_j, g_j\}$ be as in (a). For f in $End_{R_1 \otimes R_2}(P_1 \otimes P_2)$ let its image be $\sum_{i,j} \theta_j(\pi_i[f(x_j \otimes y_i)])$ where $\theta_j : P_1 \otimes End_{R_2}(P_2) \to End_{R_1}(P_1) \otimes End_{R_2}(P_2)$ by $\theta_j(a \otimes h) = f_j(\ )a \otimes h$ and $\pi_i : P_1 \otimes P_2 \to P_1 \otimes End_{R_2}(P_2)$ by $\pi_i(b_1 \otimes b_2) = b_1 \otimes g_i(\ )b_2$.

1.6 Corollary. Let S be a commutative R-algebra, P a finitely generated (resp. faithfully) projective R-module. Then $S \otimes_R P$ is a finitely generated (resp. faithfully) projective S-module. The map $S \otimes End_R(P) \to End_S(S \otimes P)$, given by $s \otimes f \to s \otimes f$, is an isomorphism.

1.7 Exercise. Prove that 1.6 is true even if S is not commutative, when the notion "Q is faithfully projective over S" is replaced by "Q is finitely generated projective and $tr_S Q = S$".

1.8 Exercise. (a) Let P be a faithfully projective R-module. Then $End_R(P)$ is a faithfully projective R-module.
(b) Let $f : R \to S$ be a homomorphism of not necessarily

commutative rings, with respect to which S is a projective left R-module. Let P be a projective left S-module. Then P is a projective left R-module.

Let A be a ring. We shall write J(A) for the Jacobson radical of A , i.e., J(A) is the intersection of the maximal left (or right) ideals of A . The following result is well-known, and will not be proved in full generality here (see e.g., BASS [2], p. 84). However, the equivalence $(2) \Leftrightarrow (3)$ is straightforward, as is the implication $(2) \Rightarrow (1)$. Moreover $(1) \Rightarrow (2)$ follows easily from 1.2 for A commutative.

1.9 **Proposition.** (Nakayama's Lemma). *Let* A *be a ring*, I *a left ideal of* A . *The following are equivalent*.
- (1) $I \subseteq J(A)$
- (2) IM = M , *with* M *a finitely generated left* A-*module*, *implies* M = 0 .
- (3) *If* N *is a submodule of a finitely generated* A-*module* M , *and* N + IM = M , *then* N = M .

1.10 **Corollary.** *Let* A *be an* R-*algebra, finitely generated as* R-*module. Then* $J(R)A \subseteq J(A)$.

Proof. If M is a finitely generated A-module with J(R)M = M, then, since M is also finitely generated over R , we have M = 0 . Hence the implication $(2) \Rightarrow (1)$ of 1.9 shows that $J(R)A \subseteq J(A)$.

1.11 **Proposition.** *Let* A *be a ring*, I *a two-sided ideal contained in* J(A) . *Let* P, Q *be finitely generated projective* A-*modules such that* $P/IP \simeq Q/IQ$ *as* A/I-*modules. Then there is an isomorphism* $P \simeq Q$ *which induces the given one of* P/IP *with* Q/IQ .

Proof. Given an isomorphism $\bar{f}$ use projectivity of P to find f such that the diagram below commutes:

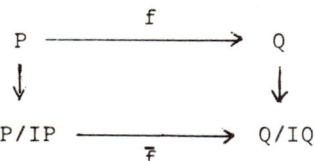

f is surjective because $\bar{f}$ is, by 1.9 . Because Q is projective, f is therefore a split epimorphism, so that $\ker(f)$ is finitely generated. But $\ker(f)$ is $0 \mod I$, hence (again by 1.9), $\ker(f) = 0$.

1.12 Corollary. *Let R be a local ring*, i.e., *R has a unique maximal ideal. Then any finitely generated projective R-module is free.*

This corollary allows us to discuss the <u>rank</u> of a finitely generated projective R-module P . For each prime ideal p of R , $P_p = R_p \otimes_R P$ is a finitely generated projective R_p-module (1.6) , hence free of rank n_p . Then $\mathrm{rank}_p P = n_p$ defines a function from $\mathrm{Spec}(R)$ to $\mathbb{Z}$. If this function is a constant n, P is said to have <u>rank n</u> .

1.13 Exercise. Let P be a finitely generated projective R-module.
  (a) Let R be an integral domain with quotient field K . Then P has rank $\dim_K(K \otimes_R P)$.
  (b) Let R be semi-local, i.e., R has only finitely many maximal ideals $m_1, \ldots, m_k$. Let P have rank n . Then P is free. (Hint: Use 1.11, the Chinese remainder theorem, and the fact

that $R_m/mR_m = R/m$.)

(c) Show that if P has rank n and admits a generating set of n elements, then that generating set is a basis (and in particular P is free).

(d) Assume R is a domain. Show that if P has rank n, then P contains a set of n linearly independent elements, and no such set with more than n elements. (In particular, P contains a free module of rank n.) Does this characterization of rank extend to more general R ?

1.14 <u>Proposition</u>. <u>Let</u> M <u>be a finitely generated</u> R-<u>module</u>, $f : M \to M$ <u>an epimorphism</u>. <u>Then</u> f <u>is an isomorphism</u>.

<u>Proof</u>. Let $S = R[f]$, a commutative subring of $\text{End}_R(M)$. The ideal $I = Sf$ satisfies $IM = M$. By 1.2, $1 - \sum_{i=1}^{n} r_i f^i$ annihilates M. Then $\sum_{i=1}^{n} r_i f^{i-1}$ is the inverse of f.

1.15 <u>Lemma</u>. <u>Let</u> A <u>be an</u> $m \times n$ <u>matrix over</u> R <u>with</u> $n \leq m$, <u>and assume there is a non-zero element</u> c <u>in</u> R <u>such that</u> $cX = 0$ <u>for all minors</u> X <u>of</u> A <u>of order</u> n. <u>Then the columns</u> $A_1, \ldots, A_n$ <u>of</u> A <u>are linearly dependent over</u> R.

<u>Proof</u>. If c also kills all minors of order $n - 1$, we are done by induction. So we may assume (renumbering the columns if necessary) that the square submatrix B of size $n - 1$ in the northwest corner of A satisfies $c(\det B) \neq 0$. Now for each $k = n, \ldots, m$ consider

$$X = \begin{pmatrix} & & & a_{1n} \\ & B & & \vdots \\ & & & a_{n-1,n} \\ a_{k1} & \cdots & & a_{kn} \end{pmatrix}$$

(where a_{ij} is the i,j entry of A). Expanding $\det X$ by the last row, and using $c(\det X) = 0$, we have

$$(*) \qquad \sum_{\ell=1}^{n-1} cX_\ell a_{k\ell} + c(\det B)a_{kn} = 0$$

where the X_ℓ are appropriate cofactors of X, for $k = n,\ldots,m$. But $(*)$ holds also for $k = 1,\ldots,n-1$, since $\det X = 0$ for such k. Hence $(*)$ holds for all $k = 1,\ldots,m$, or in other words

$$\sum_{\ell=1}^{n-1} (cX_\ell)A_\ell + c(\det B)A_n = 0 \ .$$

Since $c(\det B) \neq 0$, this is a dependence relation for the A_i, q.e.d.

1.16 <u>Exercise</u>. Let f be an endomorphism of a free R-module F of finite rank. Show:

(a) f is an isomorphism if and only if $\det f$ is a unit in R.

(b) f is surjective if and only if $\det f$ is a unit in R.

(c) f is injective if and only if $\det f$ is not a zero-divisor in R. (Hint: If f^* is the classical adjoint of f, then $f^*f = ff^*$ is multiplication by $\det f$. For the "only if" part of (c), use the case $m = n = $ rank of F of 1.15; note that f is injective if and only if the columns in a matrix for f are linearly independent.)

1.17 _Exercise._ Let $f : M \to N$ be a monomorphism of right R-modules. Let A be a projective left R-module. Then $f \otimes 1 : M \otimes_R A \to N \otimes_R A$ is a monomorphism.

The foregoing preliminaries on modules stop short of the results commonly referred to as the "Morita Theory". (For a systematic account, see BASS [1].) We will need some of these results, notably in Chapters 7 and 9, and we therefore summarize the parts we will use. In 1.18, X-Mod is the category of all left X-modules, and, when X and Y are R-algebras, an R-equivalence $F : X\text{-Mod} \to Y\text{-Mod}$ means a category equivalence with the property that the induced isomorphisms $\text{Hom}_X(M,N) \to \text{Hom}_Y(FM,FN)$ are R-linear.

1.18 Theorem. _Let_ A _be an algebra over the commutative ring_ R, Q _a faithfully projective right_ A-_module_ (_definition as in_ 1.7), _and put_ $E = \text{End}_A(Q)$. _Then_ Q _is faithfully projective as a left_ E-_module, and_ $Q \otimes_A$ _is an_ R-_equivalence_ A-Mod $\to$ E-Mod, _which restricts to an_ R-_equivalence on the categories of finitely generated projective modules._

Proof. (Leaving some details to the reader.) Define $\phi : Q \otimes_A \text{Hom}_A(Q,A) \to E$ by $(qf)x = \phi(q \otimes f)x = q(fx)$. Use 1.1 to find a projective basis for Q: there are q_i in Q and f_i in $\text{Hom}_A(Q,A)$ with $\Sigma q_i(f_i q) = q$ for all q in Q. Then ϕ is surjective: any f in E is ϕ of $\Sigma(fq_i) \otimes f_i$. Similarly, define $\psi : \text{Hom}_A(Q,A) \otimes_E Q \to A$ by $\psi(f \otimes q) = fq$; then ψ is surjective (its image is the trace ideal of Q in A). If we show that ϕ and ψ are injective as well, the statements about the category equivalence will follow: $Q \otimes_A$ and $\text{Hom}_A(Q,A) \otimes_E$ will be functors A-Mod $\to$ E-Mod (resp. E-Mod $\to$ A-Mod), and the two compositions will be isomorphic to $A \otimes_A$ and $E \otimes_E$ respectively.

Suppose $\Sigma x_j \otimes \alpha_j$ is in the kernel of ϕ (i.e., $\Sigma x_j \alpha_j = 0$) for some x_j in Q, α_j in $\text{Hom}_A(Q,A)$. Then, using $\Sigma q_i f_i = 1$ in E, we have $\Sigma x_j \otimes \alpha_j = \Sigma \Sigma x_j \otimes \alpha_j (q_i f_i) = \Sigma \Sigma x_j (\alpha_j q_i) \otimes f_i = \Sigma \Sigma (x_j \alpha_j) q_i \otimes f_i = 0$. Thus ϕ is injective. A similar argument works for ψ: given $\Sigma \beta_j \otimes y_j$ in the kernel of ψ choose g_k in $\text{Hom}_A(Q,A)$ and x_k in Q with $\Sigma g_k x_k = 1$ in A, then $\Sigma \beta_j \otimes y_j = \Sigma \Sigma \beta_j \otimes (y_j g_k) x_k = \Sigma \Sigma (\beta_j y_j) g_k \otimes x_k = 0$.

It remains to see that Q, viewed as left E-module, is faithfully projective. Define h_i in $\text{Hom}_E(Q,E)$ by $h_i q = \phi(q \otimes f_i)$; then, denoting by γ the canonical map $\text{Hom}_E(Q,E) \otimes_E Q \to E$ we have $\gamma(\Sigma h_i \otimes f q_i) = f$ for any f in E. Thus γ is surjective i.e., $\text{tr}_E(Q) = E$. Finally, Q is finitely generated and projective over E: define $\phi_k : Q \to E$ by $\phi_k(q) x = q g_k(x)$; then for any y in Q we have $y = \Sigma y(g_k x_k) = \Sigma \phi_k(y) x_k$, so that $\{\phi_k, x_k\}$ is a projective basis for Q over E.

CHAPTER 2. CENTRAL SEPARABLE ALGEBRAS.

R will be (as always) a commutative ring. For an R-algebra A, we shall let A^o denote the __opposite algebra__ of A, whose multiplication is $a^o b^o = (ba)^o$, and whose R-module structure coincides with that of A. The __enveloping algebra__ is defined by $A^e = A \otimes_R A^o$.

A^e has the following useful property: an (A,A)-bimodule M whose induced left and right R-actions coincide (i.e. $a(mb) = (am)b$ and $rm = mr$) is characterized as a left A^e-module, with action $(a \otimes b^o)x = axb$. In particular, A is a left A^e-module. There is an epimorphism of left A^e-modules

$$\phi_A : A^e \to A$$

satisfying $\phi_A(a \otimes b^o) = ab$.

2.1 __Definition__. A is said to be R-__separable__ if ϕ_A splits as an A^e-homomorphism, or equivalently, if A is A^e-projective.

If A is R-separable then $\phi = \phi_A$ is split by an A^e-map $j : A \to A^e$ ($\phi j = 1$). Thus there exists an element $e = j(1)$ in A^e satisfying $\phi(e) = 1$ and $(1 \otimes a^o)e = (a \otimes 1^o)e$ for all a in A (since $j(a) = j([1 \otimes a^o]1)$, etc.). Conversely, if such an e exists, then A is R-separable.

Writing $e = \Sigma x_i \otimes y_i^o$, the above conditions on e become that $\Sigma x_i y_i = 1$ and $\Sigma a x_i \otimes y_i^o = \Sigma x_i \otimes (y_i a)^o$. It is straightforward to compute that e is then idempotent ($e^2 = e$). We record these facts:

2.2 **Proposition.** A *is* R-*separable iff there exists an element* e *in* A^e, *necessarily idempotent, satisfying* $\phi(e) = 1$ *and* $(1 \otimes a^o)e = (a \otimes 1^o)e$ *for all* a *in* A. Such an e is called a *separability idempotent* for A.

When A is R-separable, the center of A is just eA. For $(ea)b = (1 \otimes b^o)ea = (b \otimes 1^o)ea = b(ea)$, hence $eA \subseteq$ center A. Conversely, if x is central in A then $ex = \phi(e)x = x$ (write $e = \Sigma x_i \otimes y_i^o$). More generally, if M is a left A^e-module let

$$M^A = \{x \text{ in } M \mid (1 \otimes a^o)x = (a \otimes 1^o)x \text{ for all } a \text{ in } A\}$$
$$= \{x \text{ in } M \mid xa = ax \text{ for all } a \text{ in } A\} \ .$$

The type of argument used above then yields easily:

2.3 **Proposition.** *Let* A, C *be separable* R-*algebras, having separability idempotents* e_A, e_C *respectively. Let* S *be a commutative* R-*algebra. Let* M *be an* (A,A)-*bimodule. Let* $f : A \to B$ *be an* R-*algebra homomorphism onto* B. *Then:*

(a) $M^A = e_A M$.

(b) B *is* R-*separable, with separability idempotent* $(f \otimes f^o)(e_A)$. *Moreover* $f(\text{center } A) = \text{center } B$.

(c) $D = S \otimes_R A$ *is* S-*separable. Identifying* $D \otimes_S D^o$ *with* $S \otimes_R A \otimes_R A^o$, $1 \otimes e_A$ *is a separability idempotent for* $S \otimes_R A$, *and* $\text{center}(S \otimes_R A) = S \otimes_R \text{center } A$.

(d) $E = A \otimes_R C$ *is* R-*separable. Identifying* E^e *with* $A^e \otimes_R C^e$, $e_A \otimes e_C$ *is a separability idempotent for* $A \otimes_R C$ *and* $\text{center}(A \otimes_R C) = (\text{center } A) \otimes_R (\text{center } C)$.

2.4 **Proposition.** <u>Let A be R-separable and let M be a left A-module which is R-projective. Then M is A-projective as well.</u>

Proof. Let $f : P \to M$ be an A-module epimorphism, and let $g : M \to P$ be an R-homomorphism splitting f, i.e., $fg = 1$. $\text{Hom}_R(M,P)$ is a left A^e-module via $(a \otimes b^o)h(m) = ah(bm)$. Let $\bar{g} = eg$, where e is a separability idempotent for A. A computation shows that $\bar{g}$ is an A-homomorphism and that $f\bar{g} = 1$. Thus M is A-projective.

2.5 **Examples.** The following instances of separable algebras provide counterexamples to converses of 2.4 and 2.6.

(a) $R = \mathbb{Z}$, $A = \mathbb{Q}$.
(b) $R = \mathbb{Z}$, $A = \mathbb{Z}/2\mathbb{Z}$.
(c) $R = \mathbb{Z}[\sqrt{-3}]$, $A = \mathbb{Z}[\omega]$, ω a primitive cube root of 1.

In each of these, A is R-separable, and in none of them is A R-projective. In (c), A is finitely generated and faithful over R. The verifications are left to the reader, with the following hints for part (c): A is R-separable because using the facts that 2ω is in R and $2\omega \otimes 1 = 1 \otimes 2\omega$, one can show that $1 \otimes 1 + 1 \otimes \omega - \omega \otimes 1$ is a separability idempotent. To see that A is not R-projective, use 1.4 and the fact that A and R are both free of rank 2 over $\mathbb{Z}$.

2.6 **Proposition.** <u>Let A be a separable R-algebra which is a projective R-module. Then A is a finitely generated R-module.</u>

Proof. (This result was first proved by VILLAMAYOR & ZELINSKY [1].) Let $\{x_i^o, f_i\}$ be a projective basis for A^o over R, i.e., x_i^o in A, $f_i : A^o \to R$ satisfy $x^o = \Sigma f_i(x^o) x_i^o$ for all x^o in A^o. Then $\{1 \otimes x_i^o, 1 \otimes f_i\}$ is a projective basis for $A \otimes_R A^o$ over A. Let e be a separability idempotent for A, and let $I = \{i \mid (1 \otimes f_i)(e) \neq 0\}$, a finite set.

For any x^o in A, $I \supseteq \{i \mid x(1 \otimes f_i)(e) \neq 0\} = \{i \mid (1 \otimes f_i)(x \otimes 1^o)e \neq 0\} = \{i \mid (1 \otimes f_i)([1 \otimes x^o]e) \neq 0\}$. Recall that ϕ is the multiplication map from $A \otimes_R A^o$ to A and let $e = \Sigma a_j \otimes b_j^o$. For x in A we obtain

$$x = \phi([1 \otimes x^o] e)$$

$$= \phi(\Sigma_i [(1 \otimes f_i)([1 \otimes x^o] e)](1 \otimes x_i^o))$$

$$= \phi(\sum_{i \text{ in } I} [(1 \otimes f_i)([1 \otimes x^o] e)](1 \otimes x_i^o))$$

$$= \phi(\sum_{i \text{ in } I} \sum_j [(1 \otimes f_i)(a_j \otimes (b_j x)^o)](1 \otimes x_i^o))$$

$$= \phi(\Sigma [a_j \otimes f_i((b_j x)^o)](1 \otimes x_i^o))$$

$$= \sum_{i \text{ in } I} \sum_j a_j f_i((b_j x)^o) x_i \quad .$$

I is finite and independent of x. A is thus generated by $\{a_j x_i \mid i \text{ in } I\}$.

2.7 Corollary. Let K be a field and A a separable K-algebra. Then $\dim_K A$ is finite and $L \otimes_K A$ is semisimple for every field extension L of K.

Proof. That $\dim_K A$ is finite follows from 2.6 . By 2.3(c), $L \otimes_K A$ is L-separable. Then 2.4 implies that every $L \otimes_K A$-module is projective. But semisimplicity of a ring A (i.e., A being a finite product of simple rings) is equivalent to every A-module being projective (see BASS [2], p. 79) .

2.8 Remark. Let K be a field. If A is a finite dimensional K-algebra, semisimplicity of $L \otimes_K A$ for every extension L of K is equivalent to A being K-separable. We shall not prove this. (See BASS [1], Theorem 3.2, p. 100) .

2.9 Definition. An R-algebra A that is faithful (R $\subseteq$ A) , separable over R and has R as its center, will be called a central separable R-algebra.

Suppose for a moment that R is a field and A is a central separable R-algebra. By 2.7 , A is semisimple, i.e., a finite product of simple rings A_i . The center of A being R , there can be but one A_i , i.e., A is simple and has center R . For an arbitrary commutative ring R , a central separable R-algebra need not be simple; but we shall see that its only two-sided ideals are of the form IA , with I an ideal of R .

As a first step, consider a maximal two-sided ideal $\mathcal{A}$ of A , and let I = $\mathcal{A} \cap$ R . A/$\mathcal{A}$ is a simple ring, i.e., it has no proper two-sided ideals. By 2.3(b), the center of A/$\mathcal{A}$ is R/I , and the latter must then be a field since A/$\mathcal{A}$ is simple.

Taking S = R/I , 2.3(c) implies that A/IA is a central separable R/I-algebra. The discussion above shows that A/IA is then simple. Thus IA is a maximal two-sided ideal of A , and $\mathcal{A} = (\mathcal{A} \cap R)A$.

It is a consequence of this that A is a finitely generated projective R-module. For let e be a separability idempotent for A . Then $A^e e A^e$ is a two-sided ideal of the central separable R-algebra A^e (using 2.3(d)) ; if it were a proper ideal, there would be a proper ideal I in R with $A^e e A^e \subseteq IA^e$ and IA^e proper. Applying $\phi : A^e \rightarrow A$ to this inclusion yields that A = IA , hence $A^e = IA^e$, a contradiction. Choose α_i , β_i , $i = 1,\ldots,n$ in A^e with $1 \otimes 1^o = \Sigma \alpha_i e \beta_i$. Since $e\beta_1$ is in $(A^e)^A$ (2.3(a)) , we may assume that $\alpha_i = x_i \otimes 1^o$, x_i in A . Define $f_i : A \rightarrow R$ by $f_i(x) = (e\beta_i)x$ (again 2.3(a) is being used). Then $x = (1 \otimes 1^o)x = \Sigma f_i(x) x_i$, proving that A has a finite projective basis $\{x_i, f_i\}$. Noting that left multiplication by e sends A onto R completes the proof of:

2.10 <u>Proposition</u>. Let A <u>be a central separable</u> R-<u>algebra</u>. <u>Then</u> A <u>is a finitely generated projective</u> R-<u>module</u>. <u>The inclusion map embeds</u> R <u>as a direct summand of</u> A .

2.11 <u>Corollary</u>. Let A <u>be a central separable</u> R-<u>algebra</u>. Let $\mathcal{A}$ <u>be a two-sided ideal of</u> A , <u>and</u> I <u>an ideal of</u> R . <u>Then</u> IA $\cap$ R = I <u>and</u> $\mathcal{A} = (\mathcal{A} \cap R)A$. <u>Hence</u> I $\mapsto$ IA <u>and</u> $\mathcal{A} \mapsto \mathcal{A} \cap R$ <u>yield isomorphisms</u>, <u>each the inverse of the other</u>, <u>between the lattices of ideals of</u> R <u>and two-sided ideals of</u> A .

<u>Proof</u>. That IA $\cap$ R = I follows from 2.10, i.e., from the

existence of an R-module map $t : A \to R$ satisfying $t(1) = 1$. To prove that $\mathcal{A} = (\mathcal{A} \cap R)A$ we recall the projective basis used to prove 2.10. There exist elements $\beta_1, \ldots, \beta_n$ in A^e, $x_1, \ldots, x_n$ in A such that if $f_i : A \to R$ is given by $f_i(x) = e\beta_i x$, then $x = \Sigma f_i(x)x_i$ for x in A. Then $f_i(\mathcal{A}) = (e\beta_i)\mathcal{A} \subseteq \mathcal{A}$ since $\mathcal{A}$ is an A^e-submodule of A. Hence $\mathcal{A} = \Sigma f_i(\mathcal{A})x_i \subseteq (\mathcal{A} \cap R)A \subseteq \mathcal{A}$, and the proof is done.

2.12 <u>Corollary</u>. (a) <u>Let</u> A <u>be a central separable</u> <u>R-algebra. If</u> M <u>is an</u> A-<u>module which is faithful as an</u> <u>R-module, then</u> M <u>is faithful as an</u> A-<u>module</u> (i.e., $\operatorname{ann}_A(M) = 0$).

(b) <u>Any</u> <u>R-algebra homomorphism from a central separable</u> <u>R-algebra to a faithful</u> <u>R-algebra is a monomorphism</u>.

2.13 <u>Lemma</u>. <u>Let</u> A <u>be a central separable</u> <u>R-algebra, and</u> M <u>a left</u> A^e-<u>module. Then the natural map</u> $A \otimes_R M^A \to M$, <u>given</u> <u>by</u> $a \otimes m \to am$, <u>is an isomorphism</u>.

<u>Proof</u>. (a) Maintain x_i, β_i, e as in the proof preceding the statement of 2.10: $1 \otimes 1^o = (x_i \otimes 1^o)e\beta_i$, with e, β_i in A^e. It follows at once from this formula, and from the fact that $M^A = eM$ (2.3), that the map is onto.

Suppose $z = \Sigma a_j \otimes m_j$ in $A \otimes_R M^A$ satisfies $\Sigma a_j m_j = 0$. Then $z = \Sigma a_j \otimes m_j = \Sigma[(x_i \otimes 1^o)e\beta_i]a_j \otimes m_j = \Sigma(x_i \otimes 1^o)[(e\beta_i)a_j] \otimes m_j$. But $e\beta_i a_j$ is in $eA = A^A = R$. The last expression yields $z = \Sigma x_i \otimes [(e\beta_i)a_j]m_j = \Sigma x_i \otimes (e\beta_i)(a_j m_j)$. Thus $z = 0$ and the map is injective.

2.14 **Theorem.** _Let_ A _be an_ R_-algebra. The following are equivalent conditions_:

(1) A _is a central separable_ R_-algebra._

(2) A _is faithfully projective as an_ R_-module, and the_ R_-algebra map_ $\eta_A : A^e \to \text{End}_R(A)$, _given by_ $\eta_A(a \otimes b^o)(x) = axb$, _is an isomorphism_.

Proof. (1) $\Rightarrow$ (2) : Taking $M = \text{End}_R(A)$, it follows from 2.13 that η_A is an isomorphism (the A^e-action on M is $[(a \otimes b^o)f](x) = af(bx))$.

(2) $\Rightarrow$ (1) : Let $t : A \to R$ be a splitting for the inclusion of R in A (1.4 or 2.10). It is easily checked that if $\eta_A(e) = t$ then e is a separability idempotent for A. Then $eA = tA = R$ is the center of A (2.3(a)).

2.15 Exercise. (a) Let A, B be R-algebras which are faithfully projective R-modules. Show that if $A \otimes_R B$ is central separable, so are A and B (use 1.5(b)).

(b) Let $A \subseteq B$ be an inclusion of central separable R-algebras. Show that B^A is a central separable R-algebra. If $C = B^A$ show that $B^C = A$. This result is known as the "double centralizer theorem".

2.16 Exercise. Prove the following generalization of 2.11 : Let A, B be R-algebras, with A central separable. Then $\mathfrak{b} \to A \otimes \mathfrak{b}$ is a bijection from the set of two-sided ideals of B to the set of two-sided ideals of $A \otimes B$. (Hint: Injectivity holds because A is faithfully projective; similarly for {ideals of $A \otimes B$} $\to$ {ideals of $A^e \otimes B$}. Hence there is an injection

α : {ideals of B} $\to$ {ideals of $A^e \otimes B$} . But $A^e \otimes B \simeq$ $\text{End}_R(A) \otimes B \simeq \text{End}_B(A \otimes B)$ yields a bijection β : {ideals of $\text{End}_B(A \otimes B)$} $\to$ {ideals of $A^e \otimes B$} . Now show that $A \otimes B$ being a faithfully projective right B-module (cf. (1.7)) yields a bijection γ : {ideals of B} $\to$ {ideals of $\text{End}_B(A \otimes B)$} , with $\beta\gamma = \alpha = A^e \otimes (\ )$ (cf. 1.18). Conclude that α is a bijection.

2.17 <u>Example</u>. Let P be a faithfully projective R-module, $A = \text{End}_R(P)$. Then A is a central separable R-algebra: Let $\{x_i, f_i\}$ be a projective basis of P , i.e., $\Sigma f_i(x) x_i = x$ for x in P . Let $g_j : P \to R$, y_j in P satisfy $\Sigma g_j(y_j) = 1$ (cf. 1.3) . Define E_{ij}, F_{ji} in A by $E_{ij}(x) = g_j(x) x_i$, $F_{ji}(x) = f_i(x) y_j$. Let $e = \sum_{i,j} E_{ij} \otimes F_{ji}^o$ in $A \otimes_R A^o$. We claim that e is a separability idempotent for A .

It is an easy computation that $\phi_A(e) = 1$. To check that $(f \otimes 1^o) e = (1 \otimes f^o) e$ it suffices to check for a finite set of f generating A . One such is $\{D_{k\ell}\}$, where $D_{k\ell}(x) = f_\ell(x) x_k$: for if f is in A and $f(x_k) = \Sigma c_{k\ell} x_\ell$ then $f(x) = \Sigma f(f_k(x) x_k) = \Sigma f_k(x) c_{\ell k} x_\ell = \Sigma c_{\ell k} D_{\ell k}(x)$.

Checking that $(D_{k\ell} \otimes 1^o) e = (1 \otimes D_{k\ell}^o) e$ is straightforward. Let $f_k(x_i) = r_{ik}$. Then $(D_{k\ell} \otimes 1^o) e = \sum_{i,j} D_{k\ell} E_{ij} \otimes F_{ji}^o$. But using $D_{k\ell} E_{ij} = E_{kj} r_{i\ell}$, and $\sum_i r_{i\ell} F_{ji}^o = F_{j\ell}^o$, etc., one easily checks the desired relation.

That $\text{End}_R(P)$ is central follows from 2.3(a) or by a direct computation. We leave this as an exercise.

2.18 <u>Definition</u>. We proceed to construct the <u>Brauer group</u>. For A, B central separable R-algebras, write $A \sim B$ if there exist faithfully projective R-modules P, Q such that $A \otimes_R \text{End}_R(P) \simeq B \otimes_R \text{End}_R(Q)$ (R-algebra isomorphism). By 1.5(b), $\sim$ is an equivalence relation. Let $B(R)$ denote the set of equivalence classes of central separable R-algebras. Write $[A]$ for the class of A in $B(R)$. Define a product on $B(R)$ by $[A][B] = [A \otimes_R B]$; this is well-defined by 1.5(b). It is easily checked that $B(R)$ is an abelian group, with $[A]^{-1} = [A^o]$ and $[R] = 1$ (cf. 2.14, 2.17). $B(R)$ is called the Brauer group of R.

2.19 Example. To illustrate the significance of the Brauer relation, we remark that if A and B are central separable R-algebras, then $[A] = [B]$ in $B(R)$ if, and only if, the categories A-Mod and B-Mod are R-equivalent (the terminology was introduced just before 1.18). The proof for the "if" part requires more of the Morita Theory than we have given in 1.18 — essentially, the fact that any equivalence of module categories is of the kind described there. But the "only if" part is easily proved, and we shall need this part in Chapter 9. Suppose, then, that $[A] = [B]$ in $B(R)$, i.e., $\text{End}_R(P) \otimes A$ and $\text{End}_R(Q) \otimes B$ are isomorphic R-algebras for some faithfully projective R-modules P and Q. Then by 1.7 we have that $\text{End}_A(P \otimes A)$-Mod and $\text{End}_B(Q \otimes B)$-Mod are R-equivalent, and that $P \otimes A$ (resp. $Q \otimes B$) is a faithfully projective right A- (resp. B-) module. Hence A-Mod and $\text{End}_A(P \otimes A)$-Mod are R-equivalent by 1.18, and similarly B-Mod and $\text{End}_B(Q \otimes B)$-Mod are R-equivalent, q.e.d.

2.20 Example. Let R be a commutative ring in which 2 is

invertible. The quaternions Q over R form a central separable R-algebra: Q is a free R-module with basis $1, i, j, k$ and multiplication satisfying $i^2 = j^2 = -1$, $ij = k = -ji$. That center $(Q) = R$ is clear. Let $e = \frac{1}{4}(1\otimes 1^o - i\otimes i^o - j\otimes j^o - k\otimes k^o)$. It is easily checked that e is a separability idempotent in $Q\otimes_R Q^o$ (check the conditions of 2.2 for $a = 1, i, j, k$) . $[Q]$ is an element of order ≤ 2 in $B(R)$, as $(a+bi+cj+dk) \mapsto (a-bi-cj-dk)^o$ gives an isomorphism $Q \simeq Q^o$.

2.21 **Exercise.** (a) Let 2 be invertible in R . For a,b **units** in R define the generalized quaternion algebra $\left(\frac{a,b}{R}\right)$ to be $R\oplus Ri\oplus Rj\oplus Rk$ where $i^2 = a$, $j^2 = b$, $ij = -ji = k$. Show that $\left(\frac{a,b}{R}\right)$ is a central separable R-algebra.

(b) Show that $x+yi+zj+wk \mapsto (x-yi-zj-wk)^o$ is an isomorphism $\left(\frac{a,b}{R}\right) \to \left(\frac{a,b}{R}\right)^o$; conclude that the element $[a,b]$ of $B(R)$ represented by $\left(\frac{a,b}{R}\right)$ has order ≤ 2 .

For the remainder of this exercise, R is a field of characteristic $\neq 2$, and a, b, c in R are non-zero.

(c) Suppose the equation $ax^2 + by^2 = 1$ has a solution (x,y) in R . Show that $\left(\frac{a,b}{R}\right) \simeq (R)_2$, the ring of 2×2 matrices over R . (Hint: The matrix corresponding to i is similar to a matrix $\begin{pmatrix} c & 0 \\ 0 & c \end{pmatrix}$ or $\begin{pmatrix} 0 & a \\ 1 & 0 \end{pmatrix}$. Use facts about i and j to find the matrix corresponding to j .) Conversely, show that if $\left(\frac{a,b}{R}\right) \simeq (R)_2$ then the equation above has a solution (x,y) .

(d) Show that $\left(\frac{a,a}{R}\right) \simeq (R)_2$ if and only if a is a sum of two squares in R . Show that if $\left(\frac{-1,-1}{R}\right)$ is a division ring, R has characteristic 0 .

(e) Show that $\left(\frac{a,b}{R}\right) \otimes \left(\frac{a,c}{R}\right) \simeq \left(\frac{a,bc}{R}\right) \otimes \left(\frac{1,-1}{R}\right)$ as R-algebras.

(Outline: Let $B = \{1, i, j, k\}$ (resp. $B' = \{1, i', j', k'\}$) be a standard basis for $\left(\frac{a,b}{R}\right)$ (resp. $\left(\frac{a,c}{R}\right)$). The sixteen elements $x \otimes y$ (x in B, y in B') are a basis for $\left(\frac{a,b}{R}\right) \otimes \left(\frac{a,c}{R}\right)$. Let X (resp. Y) be the linear span of $\{1 \otimes 1, i \otimes 1, j \otimes j', k \otimes j'\}$ (resp. $\{1 \otimes 1, 1 \otimes j', i \otimes k', -c(i \otimes i')\}$) in $\left(\frac{a,b}{R}\right) \otimes \left(\frac{a,c}{R}\right)$. Show that $X \simeq \left(\frac{a,bc}{R}\right)$ and $Y \simeq \left(\frac{c,-a^2c}{R}\right) \simeq \left(\frac{1,-1}{R}\right)$ and that X and Y commute elementwise (check this on the given bases). By the universal property of $\otimes_R$, the inclusions of X and Y in $\left(\frac{a,b}{R}\right) \otimes \left(\frac{a,c}{R}\right)$ therefore induce a homomorphism $\left(\frac{a,bc}{R}\right) \otimes \left(\frac{1,-1}{R}\right) \to \left(\frac{a,b}{R}\right) \otimes \left(\frac{a,c}{R}\right)$. For dimension reasons, this is an isomorphism.)

(f) As above, let $[a,b]$ be the class in $B(R)$ of $\left(\frac{a,b}{R}\right)$. Show that $[a,b][a,c] = [a,bc]$. Let $Quat(R)$ denote the subgroup of $B(R)$ generated by the $[a,b]$, and let $_2B(R)$ denote the group of elements of order ≤ 2 in $B(R)$. Show that $Quat(R) \subseteq {}_2B(R)$ (this was done in part (b); the relation $[a,b][a,c] = [a,bc]$, with part (c), gives a second proof). Is the reverse inclusion $_2B(R) \subseteq Quat(R)$ valid? (Undying fame awaits the solver of this last part.)

(g) The above considerations show that $a,b \mapsto [a,b]$ is a symmetric bilinear pairing $(R^{\cdot}/R^{\cdot 2}) \times (R^{\cdot}/R^{\cdot 2}) \to Quat(R)$. Call R <u>pythagorean</u> if, given x, y in R, there exists z in R with $x^2 + y^2 = z^2$. Show that the pairing is nondegenerate ($[a,x]$ trivial in $B(R)$ for all x implies a is a square) if R is pythagorean. The pairing can, in general, fail to be nondegenerate. It is an open problem to characterize fields for which it is nondegenerate (see KAPLANSKY [1]).

2.22 **Exercise**. (F. DeMeyer) Let B be a commutative separable subalgebra of a projective separable R-algebra A. Show B is projective.

CHAPTER 3.   THE BRAUER GROUP OF A FIELD.

Our point of view in these notes is that the classical theory of central simple algebras over fields is more or less "known". We include here, primarily for motivation, an indication of how that theory goes. Good references are Chapter 4 of HERSTEIN [1], SERRE [1] and Chapter 8 of ARTIN, NESBITT & THRALL [1] . We give proofs here only when they involve a conceptually important method and don't take us too far afield.

If A is a central separable algebra over a field K , we have seen that A is central <u>simple</u> (discussion following 2.9) and finite-dimensional (2.6) . Conversely, if A is finite-dimensional and central simple over K , the map $A \otimes A^o \to \text{End}_K(A)$ is an isomorphism and A is therefore central separable (see Theorem 4.13 of HERSTEIN [1]). Thus, over fields, <u>central separable algebras</u> are the same as <u>finite-dimensional central simple algebras</u>. We will use the term <u>central separable</u>; but note that the traditional nomenclature is <u>central simple</u>, with finite dimensionality tacitly understood.

Let A be a central separable K-algebra where K is a field. Wedderburn's theorem asserts that A is isomorphic, as a K-algebra, to a matrix ring $(D)_n$ for some central K-division algebra D . Moreover, D is unique up to isomorphism: $(D_1)_{n_1} \simeq (D_2)_{n_2}$ implies $D_1 \simeq D_2$ and $n_1 = n_2$. D is called the <u>division ring component</u> of A . It is an easy exercise to show that if A_1 and A_2 are central separable K-algebras with division ring components D_1 and D_2 respectively, then $[A_1] = [A_2]$ in $B(K)$ if and only if $D_1 \simeq D_2$ as K-algebras.

3.1 __Example.__ __If K is an algebraically closed field then__ $B(K) = 0$. For suppose D is a finite-dimensional central division algebra over K . Let α be in D ; then $K[\alpha] = K(\alpha)$ is a finite algebraic extension field of K , hence $K(\alpha) = K$; done.

We will prove in Chapter 8 that when K is algebraically closed, $B(K(X))$ is also 0 . A description, for arbitrary fields, of $B(K(X))$ in terms of $B(K)$ and the arithmetic of $K(X)$, is an important open problem. (See §4 of AUSLANDER & BRUMER [1].)

3.2 __Corollary.__ __If A is a central separable algebra over a field K then__ $\dim_K A$ __is a square__. For let $\bar{K}$ be an algebraic closure of K , and put $\bar{A} = \bar{K} \otimes_K A$; then $\dim_K A = \dim_{\bar{K}} \bar{A} = \dim_{\bar{K}}(\bar{K})_n = n^2$.

In fact the same argument shows that if A is a central separable R-algebra over any commutative ring R , then $\text{rank}_p A$ (definition precedes 1.13) is a square for each prime ideal p . (Proof: $\text{rank}_p A$ equals the dimension of $(R_p/pR_p) \otimes A$ over the field R_p/pR_p .)

Let A be a central separable algebra over the field K , and let D be its division ring component. Then $\dim_K D$ is a square by 3.2 . The integer $\delta(A) = \sqrt{\dim_K D}$ is called the __index__ of A , or of [A] . (Thus, for example, [A] is trivial in $B(K)$ iff $\delta(A) = 1$.) The uniqueness part of Wedderburn's theorem guarantees that $\delta(A)$ and $\delta[A]$ are well-defined, and yields:

3.3 Proposition. *If* L *is a finite extension field of* K *and* $L \otimes_K A$ *is trivial in* $B(L)$ *then* $\delta(A)$ *divides* $[L:K]$.

Proof. We may assume $A = D$ is a central division algebra. Say $\dim_K D = \delta^2$, $[L:K] = n$, and $L \otimes_K D \simeq (L)_m$. Counting dimensions shows that $\delta = m$; we must show $\delta | n$. Since each element of L is (by multiplication) a K-endomorphism of L, a choice of basis for L over K gives an embedding $L \to (K)_n$. Hence $(K)_\delta \subseteq (L)_\delta \simeq L \otimes_K D \subseteq (K)_n \otimes_K D \simeq (D)_n$. By the double-centralizer theorem (2.13 and 2.15(b)) it follows that $(K)_\delta \otimes B \simeq (D)_n$ with B central separable. Let Δ be the division ring component of B, say $B \simeq (\Delta)_t$; then $(D)_n \simeq (K)_\delta \otimes B \simeq (B)_\delta \simeq (\Delta)_{t\delta}$ shows that $D \simeq \Delta$ and $t\delta = n$.

The index $\delta(A)$ has special significance when A is a division algebra:

3.4 Theorem. *If* A *is a finite-dimensional central division algebra over the field* K *and* L *is a maximal subfield of* A *then* $[L:K] = \delta(A)$, *i.e.*, $\dim_K A = [L:K]^2$.

For a proof see Theorem 4.2.2 of HERSTEIN [1]. Note that 3.4 is false if "central division algebra" is replaced by "central simple algebra": take $K = \mathbb{C}$, $A = (\mathbb{C})_n$ for a counterexample. The significance of maximal subfields of central separable algebras will become clearer shortly (see 3.14 ff. below).

The Skolem-Noether theorem, 3.6 below, is a key tool in studying central separable algebras over fields. It will give us two additional examples (3.7 and 3.8).

3.5 **Lemma.** Let C be a finite-dimensional simple K-algebra, K a field, and let M and N be C-modules. If $\dim_K M = \dim_K N$ then $M \simeq N$ (as C-modules).

Proof. C is, among other things, a simple Artinian ring. The structure theory for such rings tells us that C has a unique (up to isomorphism) indecomposable module X, and that any C-module is a direct sum of copies of X. The assertion of the lemma is then obvious.

3.6 **Theorem.** (Skolem-Noether). Let $B \subseteq A$ be K-algebras (K a field) with B simple and A central separable. Let $f : B \to A$ be a K-algebra homomorphism. Then f extends to an inner automorphism of A: there is a unit u in A such that $f(b) = ubu^{-1}$ for all b in B.

Proof. Let L be the field center(B). Then $B \otimes_K A^o$ is a central separable L-algebra (see 2.3(d)), and in particular a simple ring. A is a $B \otimes_K A^o$-module via $(b \otimes a^o)x = bxa$. Let A_1 be A as K-vector space, viewed as a $B \otimes_K A^o$-module via $(b \otimes a^o)x = f(b)xa$. By 3.5, A and A_1 are isomorphic $B \otimes A^o$-modules. Choose a $B \otimes A^o$-isomorphism $h : A \to A_1$, and define u and v in A by $h(1) = u$, $h(v) = 1$. We will show: (a) $uv = vu = 1$, and (b) $ub = f(b)u$ for all b in B. Since (a) says precisely that u is a unit, this will complete the proof.

For (a), we have $1 = h(1 \cdot v) = h(1 \otimes v^o \cdot 1) = (1 \otimes v^o)h(1) = h(1)v = uv$ and $h(1) = u = 1 \otimes u^o \cdot 1 = (1 \otimes u^o)h(v) = h(1 \otimes u^o \cdot v) = h(vu)$. Since h is injective, this proves $uv = vu = 1$.

For (b), we have (for any b in B) $ub = h(1)b = (1 \otimes b^o)h(1) = h(1 \otimes b^o \cdot 1) = h(b) = h(b \otimes 1^o \cdot 1) = (b \otimes 1^o)h(1) = f(b)h(1) = f(b)u$.

3.7 Example. If K is a finite field then $B(K) = 0$.
This boils down rapidly to another "Wedderburn's theorem": a finite division ring is a field. As such it can be proved directly (see Theorem 3.1.1 of HERSTEIN [1]). Alternatively, one can use 3.4 and the Skolem-Noether theorem to reduce the proof to an exercise about finite groups. For this approach see HERSTEIN [1], p. 102, example (2) .

3.8 Example. If $K = \mathbb{R}$ then $B(K) = \mathbb{Z}/2\mathbb{Z}$.

Proof. The ordinary quaternions $\mathcal{H} = \left(\frac{-1,-1}{\mathbb{R}}\right)$ are a finite-dimensional central division algebra over $\mathbb{R}$ (see 2.21). The claim is that $\mathcal{H}$, and $\mathbb{R}$ itself, are (up to isomorphism) the only such. 3.4 shows that any finite-dimensional central $\mathbb{R}$-division algebra D distinct from $\mathbb{R}$ has dimension 4 and contains $\mathbb{C} = \mathbb{R}(i)$ as a maximal subfield. The Skolem-Noether theorem implies that the conjugation in $\mathbb{C}$ extends to an inner automorphism, say by u , of D . One then lets j be a suitable multiple of u , puts $k = ij$, and shows that $\{1, i, j, k\}$ are linearly independent and satisfy the right relations, so that $D \simeq \mathcal{H}$. The details are left as an exercise; or see HERSTEIN [1], p. 102, example (3) .

3.9 Theorem. Let D be a finite-dimensional central K-division algebra. Then D has a maximal subfield which is a separable field extension of K .

The proof of 3.9 is slightly involved; see for example Theorem 4.3.3 of HERSTEIN [1]. It is an important result; for example, we will see below that it is a key step in proving that $B(K)$ is a torsion group. One immediate application of 3.9 is to sharpen 3.1 :

3.10 Example. *If K is a separably closed field then* $B(K) = 0$. For if D is a finite-dimensional central K-division algebra, the maximal subfield whose existence is given by 3.9 must be K, and then $D = K$ by 3.4.

About 1930, Brauer, Hasse, Noether, Albert, and others succeeded in computing $B(K)$ for number fields (more generally, global fields) and their completions (more generally, local fields). These theorems are much deeper than the examples above, and they are the crowning glory of the classical theory. The results for $K = \mathbb{Q}$ are cited here; the generalizations are easily guessed.

3.11 Theorem. *If p is a prime integer and K is the p-adic completion $\mathbb{Q}_p$ of $\mathbb{Q}$ then* $B(K) \simeq \mathbb{Q}/\mathbb{Z}$.

This theorem, or rather the more general one where K is an arbitrary local field, is a central result in local class field theory.

3.12 Theorem. *If x is in $B(\mathbb{Q})$ then the image x_p of x in $B(\mathbb{Q}_p)$ (where $B(\mathbb{Q}) \to B(\mathbb{Q}_p)$ is induced by the inclusion of $\mathbb{Q}$ in $\mathbb{Q}_p$) is nontrivial for only finitely many p. Hence the induced map $B(\mathbb{Q}) \to \prod_p B(\mathbb{Q}_p)$ lands in $\bigoplus_p B(\mathbb{Q}_p)$. Moreover,*

$B(\mathbb{Q}) \to \underset{p}{\oplus} B(\mathbb{Q}_p)$ is injective, and the image is precisely the set of (x_p) such that $\underset{p}{\Sigma} x_p = 0$ in $\mathbb{Q}/\mathbb{Z}$. In other words,

$$0 \to B(\mathbb{Q}) \to \underset{p}{\oplus} B(\mathbb{Q}_p) \overset{\text{sum}}{\to} \mathbb{Q}/\mathbb{Z} \to 0$$

is exact.

This theorem, or rather the more general one where $\mathbb{Q}$ is replaced by an arbitrary global field, is a central result in global class field theory.

In the symbols $\underset{p}{\Pi}$, $\underset{p}{\oplus}$ and $\underset{p}{\Sigma}$ occurring in 3.12, p runs over all prime integers and $p = \infty$, where as usual $\mathbb{Q}_\infty$ means $\mathbb{R}$, and $B(\mathbb{R})$ is embedded as the subgroup $\{0, \frac{1}{2}\}$ of the corresponding copy of $\mathbb{Q}/\mathbb{Z}$. 3.11 - 3.12 say that each finite-dimensional central division algebra over $\mathbb{Q}$ can be specified uniquely by a finite set of rational numbers between 0 and 1, in such a way that (1) a set X of rationals corresponds to an element of $B(\mathbb{Q})$ if and only if $\underset{i \text{ in } X}{\Sigma} i$ is in $\mathbb{Z}$, and (2) the product in $B(\mathbb{Q})$ corresponds to addition ("componentwise" — the components indexed by p) of rationals modulo 1.

We introduce at this point a notation which will be useful throughout.

3.13 Definition. If $f : R \to S$ is a homomorphism of commutative rings, $B(S/R)$ denotes the kernel of $B(f) : B(R) \to B(S)$. If x is in $B(S/R)$ we say S splits x, or S is a splitting ring for x; if A is a central separable R-algebra with $[A] = x$ we also say S splits A. The notation B(S/R) is abusive, since the kernel depends on f (i.e., on the way in which S is being viewed as R-algebra) and not only on R and S . In practice, f is always clear from the context; in the present chapter, f will always be the inclusion of one field in another.

Splitting rings always exist: let m be any maximal ideal of R and consider the composition $R \to R/m \to \overline{R/m}$. A major problem in the theory is to find <u>good</u> splitting rings. For fields, one has:

3.14 <u>Theorem</u>. <u>If</u> L <u>is a maximal subfield of a finite-dimensional central</u> K-<u>division algebra</u> D , <u>then</u> L <u>splits</u> D , i.e., [D] <u>is in</u> B(L/K) . (See HERSTEIN [1], Corollary on page 96.)

3.15 <u>Corollary</u>. <u>Any central separable algebra</u> A <u>over a field</u> K <u>is split by some finite Galois extension of</u> K .

<u>Proof</u>. An extension L of K splits A iff it splits the division ring component of A ; so assume A is a division ring. Then 3.9 and 3.14 conspire to show that A is split by a finite separable extension L' of K . Let L be a minimal Galois extension of K containing L' , i.e., L is the compositum (in an algebraic closure of K) of the conjugates of L' . Then L is a finite Galois extension of K which splits A ; done.

3.16 <u>Corollary</u>. <u>For any field</u> K <u>let</u> Ω(K) <u>be the set</u>

of finite Galois field extensions of K.[*] Then
$$B(K) = \bigcup_{L \text{ in } \Omega(K)} B(L/K) \ .$$

We want finally to sketch the classical cohomological description of $B(K)$, and the resulting proof that $B(K)$ is a torsion group, when K is a field. <u>By 3.16, it suffices for the latter to show that each</u> $B(L/K)$, L <u>in</u> $\Omega(K)$, <u>is a torsion group</u>. The point is that a crucial role is played in this proof by the existence of good splitting rings ("good" meaning, here, a finite Galois extension). This should be kept in mind later when we prove in Chapter 12 that $B(R)$ is torsion for <u>any</u> commutative ring R.

3.17 <u>Definition</u>. Let G be a group and let U be an abelian group written multiplicatively, on which G acts. A <u>cocycle</u> (more properly, a "2-cocycle on G in U") is a map $f : G \times G \to U$ such that $\sigma_1 f(\sigma_2,\sigma_3) f(\sigma_1,\sigma_2\sigma_3) = f(\sigma_1\sigma_2,\sigma_3) f(\sigma_1,\sigma_2)$ for all $\sigma_1, \sigma_2, \sigma_3$ in G. A <u>coboundary</u> ("2-coboundary on G in U") is a map $f : G \times G \to U$ such that $f(\sigma_1,\sigma_2)(g\sigma_1)\sigma_1(g\sigma_2) = g(\sigma_1\sigma_2)$ for all σ_1, σ_2 in G, for some (set-) map $g : G \to U$.

3.18 <u>Proposition</u>. (a) Z = {cocycles} <u>and</u> B = {coboundaries} <u>are groups</u> (<u>under</u> $(fg)(\sigma_1,\sigma_2) = f(\sigma_1,\sigma_2)g(\sigma_1,\sigma_2)$) <u>and</u> B <u>is a subgroup of</u> Z.

Use (a) to define the group $H^2(G,U) = Z/B$. Then:

---

[*] If the reader is sufficiently meticulous to worry about whether $\Omega(K)$ is a set, he may replace it by the set (!) of <u>isomorphism classes of</u> Galois extensions.

(b) If G is finite of order n, then $nH^2(G,U) = 0$, i.e., for any cocycle f the map $(\sigma,\tau) \to (f(\sigma,\tau))^n$ is a coboundary. In particular, $H^2(G,U)$ is a torsion group.

For explicit proofs of 3.18 see HERSTEIN [1], page 117-118.

The classical proof that $B(K)$ is torsion is achieved by establishing, for any finite Galois field extension L of K, an isomorphism $H^2(G,L^{\cdot}) \simeq B(L/K)$, where $G = \mathrm{Gal}(L/K)$ is the Galois group. Let $f : G \times G \to L^{\cdot}$ be a cocycle, let $\{u_\sigma\}$ be a set of symbols indexed by G, and let $A = \bigoplus_{\sigma \text{ in } G} L u_\sigma$ as abelian group. One defines a multiplication on A by putting

$$au_\sigma \cdot bu_\tau = a(\sigma b)f(\sigma,\tau)u_{\sigma\tau} \qquad (a, b \text{ in } L) .$$

Then A becomes a K-algebra (1 is $f(1,1)^{-1}u_1$, and the cocycle condition makes the multiplication associative), called the crossed product of L and G with respect to f.

3.19 Theorem. For each cocycle f, the crossed product constructed above is a central separable K-algebra split by L. The resulting map $Z \to B(L/K)$ induces a map $H^2(G,L^{\cdot}) \to B(L/K)$ which is (1) well-defined, (2) a homomorphism and in fact (3) an isomorphism.

(For proofs see HERSTEIN [1], Theorems 4.4.1 and 4.4.3 and Lemma 4.4.1 .)

We will prove a generalization (7.12 in Chapter 7) of 3.19, to the case where L/K is a Galois extension of rings.

Given a central separable algebra A over a field K, we now know that the order $e(A)$ of $[A]$ in $B(K)$ is finite.

It is remarkable that $e(A)$ turns out to be closely related to the index $\delta(A)$ (defined just before 3.3); $e(A)$ divides $\delta(A)$, and "conversely" every prime which divides $\delta(A)$ divides $e(A)$. (Thus e and δ are related in the same way as the minimal and characteristic polynomials of a matrix.) For proofs of these facts see HERSTEIN [1], pp. 119-120.

The proof of 3.8 may be imitated to yield a bit more; we leave the details to the reader: Let K be a field of characteristic $\neq 2$. Then the elements $[a,b]$ of $B(K)$ (see 2.21) are precisely the elements of index ≤ 2. The results quoted above show that x in $B(K)$ has $e(x)$ a power of 2 iff $\delta(x)$ is a power of 2, and the question raised in 2.21(f) can be rephrased as follows: does $e(x) = 2$ imply that x is a product of elements of index 2 ? Some interesting examples in this connection are to be found in early papers of A.A. Albert, e.g. ALBERT [14] and [30] of the bibliography of DEURING [2].

3.20 <u>Exercise</u>. The computation of $H^2(G,L^\cdot)$ is easier when G is cyclic. Let L/K be a Galois extension whose Galois group G is cyclic of order n generated by σ. For x in L, write $N_{L/K}(x)$ for the norm $\prod_{i=1}^{n} \sigma^i x$ of x ; note that N is a homomorphism, $L^\cdot \to K^\cdot$. Verify that the following prescriptions yield an isomorphism $H^2(G,L^\cdot) \simeq K^\cdot/N_{L/K}(L^\cdot)$: to a cocycle $f : G \times G \to L^\cdot$, assign the class of $\prod_{i=1}^{n} f(\sigma^i, \sigma)$; to the class of an element a in $K^\cdot$ assign the map (cocycle!)

$$f(\sigma^i, \sigma^j) = \begin{cases} 1 & \text{if } i + j < n \\ a & \text{if } i + j \geq n \end{cases}$$

CHAPTER 4.   LEMMAS, MOSTLY ABOUT SEPARABILITY.

The main facts about separability which we are after are <u>roughly</u> that it is "transitive" (4.4) , that it can be checked modulo maximal ideals (4.12) , and that for fields it means what it should (4.7) .

4.1 <u>Theorem</u>.(Cayley-Hamilton) <u>Let</u> R <u>be any commutative ring and let</u> $a = (a_{ij})$ <u>be in</u> $(R)_n$. <u>Let</u> $f(T) = \det(TI_n - a)$ <u>in</u> $R[T]$ <u>be the characteristic polynomial of</u> a . <u>Then</u> $f(a) = 0$ <u>in</u> $(R)_n$.

<u>Proof</u>. (L. Roberts)   Consider the commutative diagram

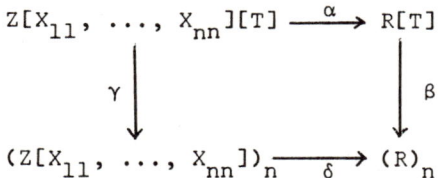

where α and δ are given by $X_{ij} \mapsto a_{ij}$, $T \mapsto T$; and β (resp. γ) is evaluation at $T = a$ (resp. at $T = $ the matrix (X_{ij})). Let $F(T)$ in $Z[X_{11}, \ldots, X_{nn}][T]$ be the characteristic polynomial of (X_{ij}) . Then $\alpha F(T) = f(T)$ and $f(a) = \beta f(T) = \delta \gamma F(T)$. But $\gamma F(T) = 0$ by the traditional Cayley-Hamilton theorem, since $Z[X_{11}, \ldots, X_{nn}]$ embeds (in many ways) in $\mathbb{C}$; done!

4.2 <u>Lemma</u>.   <u>Let</u> $R \subseteq S$ <u>be commutative rings with</u> S <u>finitely generated as</u> R-<u>module</u>. <u>If</u> $\mathcal{M}$ <u>is a maximal ideal of</u> S , $\mathcal{M} \cap R$ <u>is a maximal ideal of</u> R .

<u>Proof</u>.   Let $\mathcal{M} \cap R = \mathfrak{m}$; then the composition $R \hookrightarrow S \to S/\mathcal{M}$ has kernel $\mathfrak{m}$, and hence induces an inclusion of $R/\mathfrak{m}$ into the field $S/\mathcal{M}$. So we need:

4.2' **Lemma.** *Let* $R \subseteq S$ *with* S *a field, finitely generated as* R-*module. Then* R *is a field*.

Proof. Let $0 \neq x$ in R and let $y = x^{-1}$ in S; we show y lies in R. Because S is finitely generated over R, y is *integral* over R: there exist $n \geq 1$ and a_i in R with $y^n + a_{n-1} y^{n-1} + \ldots + a_0 = 0$. Multiplying through by x^{n-1} shows that $y = y^n x^{n-1}$ lies in R.

4.3 **Proposition.** *Let* $f: R \to S$ *be a commutative* R-*algebra, and assume that* R *is local, say with maximal ideal* $\mathfrak{m}$, *and that* S *is a finitely generated* R-*module (via* f*). Then:*

(a) *Every maximal ideal of* S *contains* $(f\mathfrak{m})S$. *(Consequently, if* $(f\mathfrak{m})S$ *is maximal,* S *is local.)*

(b) *In any case,* S *is semilocal.*

Proof. Let $R' = f(R)$. It is an easy exercise to see that R' is local, with maximal ideal $f(\mathfrak{m})$. Hence (replacing R by R') we can assume $R \subseteq S$ and write $\mathfrak{m}S$ instead of $(f\mathfrak{m})S$ in (a).[*] Part (a) now follows immediately from 4.2; moreover, for (b) it now suffices to show that $S/\mathfrak{m}S$ is semi-local. But $S/\mathfrak{m}S$ is a finite-dimensional extension of $R/\mathfrak{m}$, hence an Artinian ring. Suppose $M_1, M_2, \ldots$ are distinct maximal ideals of $S/\mathfrak{m}S$. Let $N_n = \prod_{i=1}^{n} M_i$. Then $N_k = N_{k+1} = \ldots$ for some k, and $M_{k+1} N_k = N_k$, hence $\prod_{i=1}^{k} M_i \subseteq M_{k+1}$ and $M_{k+1} = M_i$ for some $i \leq k$. Thus $S/\mathfrak{m}S$ is semi-local.

---

[*] The reader may prefer to forego the exercise and be content with the special case of 4.3 in which f is injective. Little will be lost if this alternative is chosen.

4.4 Lemma. *Let* A *be an* S*-algebra,* S *a commutative* R*-algebra, and assume* A *is* S*-separable and* S *is* R*-separable. Then* A *is* R*-separable.* (For converses, see exercise 4.14.)

Proof. Write A_R^e for $A \otimes_R A^o$ and A_S^e for $A \otimes_S A^o$. We have that S is S^e-projective, hence by 1.7, $S \otimes_{S^e} A_R^e$ is $S^e \otimes_{S^e} A_R^e = A_R^e$-projective. Now $S \otimes_{S^e} A_R^e = S \otimes_{S^e} (A \otimes_R A^o)$ is isomorphic as A_R^e-module to A_S^e; e.g. inverse isomorphisms are given by $s \otimes a \otimes a^o \mapsto sa \otimes a^o = a \otimes sa^o$ and $a \otimes a^o \mapsto 1 \otimes a \otimes a^o$. Hence A_S^e is A_R^e-projective. But we have also that A is A_S^e-projective, and therefore by 1.8(b) A is A_R^e-projective.

Exercise. Find a computational proof of 4.4, i.e., given separability idempotents for A over S and for S over R, find one for A over R.

4.5 Corollary. *If* A *and* B *are separable* R*-algebras, the direct product* $A \times B$ *is a separable* R*-algebra.*

Proof. If e_A, e_B are separability idempotents (2.2) for A, B respectively, the element (e_A, e_B) of $(A \otimes_R A^o) \times (B \otimes_R B^o) \simeq (A \times B) \otimes_{R \times R} (A \times B)^o$ is a separability idempotent for $A \times B$ over $R \times R$, so that $A \times B$ is $R \times R$-separable. (The verification of this, and of the isomorphism $A_R^e \times B_R^e \simeq (A \times B)_{R \times R}^e$ involved, are left as easy exercises.) It then suffices, by 4.4, to show that $R \times R$ is R-separable. But (exercise) $e \otimes e + f \otimes f$ is a separability idempotent for $R \times R$ (where e, f are the canonical idempotents $(1,0)$ and $(0,1)$ of $R \times R$), q.e.d.

Exercise: (a) $R \times \ldots \times R$ (finitely many copies of R) is R-separable.

(b) $R \times R \times \ldots$ (infinitely many copies of R) is not R-separable (cf. 2.6).

4.6 **Lemma**. *Let* S *and* A *be R-algebras and assume that* S *is commutative and contains* R *as a direct summand. If* $S \otimes_R A$ *is S-separable*, A *is R-separable*.

Proof. Use the direct summand hypothesis to choose t in $\text{Hom}_R(S,R)$ with $t(1) = 1$, and identify $S \otimes_R A_R^e$ with $(S \otimes_R A)_S^e$. Let $\bar{e}$ be a separability idempotent for $S \otimes_R A$; hence $\bar{e}$ is in $S \otimes_R A_R^e$. We claim that the element $e = (t \otimes 1)\bar{e}$ of A_R^e is a separability idempotent for A. There are two things to check: $(1 \otimes x^o)e = (x \otimes 1^o)e$ for all x in A, and $\phi_A(e) = 1$ (where ϕ_A is as in 2.1). For the first of these, note that
$(1 \otimes x^o)e = (t \otimes 1)[(1 \otimes 1 \otimes x^o)\bar{e}] = (t \otimes 1)[(1 \otimes x \otimes 1^o)\bar{e}] = (x \otimes 1^o)e$.
The second follows similarly from $\phi_{S \otimes A}(\bar{e}) = 1$.

4.7 **Corollary**. *Let* L *be a finite separable field extension of* K. *Then* L *is a separable K-algebra*.

(The converse is true but we will not need it: if $L \supseteq K$ are fields and L is a separable K-algebra, then L is a finite separable field extension of K. See BASS [1], Chapter III §3.)

Proof. L is of the form $K[X]/(f(X))$, and the roots of the irreducible polynomial f (in an algebraic closure of K) are distinct. Choose a splitting field E for $f(X)$ over K. Then, since f splits into distinct linear factors in $E[X]$, we have $L \otimes_K E \simeq E \times \ldots \times E$. Hence $L \otimes_K E$ is E-separable by 4.5, and consequently L is K-separable by 4.6.

Let A be an R-algebra. Define $J = J_R(A)$ by the exact sequence

$$E(A): 0 \longrightarrow J \longrightarrow A^e \xrightarrow{\phi_A} A \longrightarrow 0$$

of A^e-modules. (By definition, A is R-separable iff $E(A)$ splits.) For any left A^e-module (= "A-A bimodule over R") M, we consider the sequence $\text{Hom}_{A^e}(E(A),M)$:

$$0 \to \text{Hom}_{A^e}(A,M) \to \text{Hom}_{A^e}(A^e,M) \to \text{Hom}_{A^e}(J,M) \to 0 \, .$$

It is exact if we omit the right-hand zero, and the middle term is (naturally isomorphic to) M. We want to identify the other two terms. (The reason for this excursion is that we need 4.11 in order to prove 4.12 !)

Define $\delta: A \to J$ by $\delta a = a \otimes 1^o - 1 \otimes a^o$.

(Exercise. δ is R-linear, but not A^e-linear in general.)

4.8 Lemma. (a) The image of δ generates J as a left ideal of A^e, and

(b) $\delta(ab) = (\delta a)b + a(\delta b)$ for all a, b in A.

Proof. If $x = \Sigma a_i \otimes b_i^o$ is in J we have $\Sigma a_i b_i = 0$, hence $x = \Sigma a_i \otimes b_i^o - \Sigma a_i b_i \otimes 1^o = -\Sigma(a_i \otimes 1)\delta b_i$, which proves (a). For (b), we have $(\delta a)b + a(\delta b) = (1 \otimes b^o)(a \otimes 1^o - 1 \otimes a^o) + (a \otimes 1^o)(b \otimes 1^o - 1 \otimes b^o) = ab \otimes 1^o - 1 \otimes (ab)^o = \delta(ab)$.

4.9 Corollary. For any left A^e-module M, the canonical isomorphism $\text{Hom}_{A^e}(A^e,M) \cong M$ restricts to an isomorphism $\text{Hom}_{A^e}(A,M) \cong M^A$.

Proof. Since $A \cong A^e/J$ as A^e-modules (because $E(A)$ is exact) we have $\text{Hom}_{A^e}(A,M) = \{f \text{ in } \text{Hom}_{A^e}(A^e,M) \mid f(J) = 0\}$. The isomorphism $M \to \text{Hom}_{A^e}(A^e,M)$ in question takes x in M to "right multiplication by x", hence induces $\text{Hom}_{A^e}(A,M) \simeq \{x \in M \mid Jx = 0\}$. By 4.8(a), this is the same as $\{x \in M \mid (\delta a)x = 0, \text{ all } a \text{ in } A\}$. But this is precisely M^A (cf. the definition, just before 2.3), q.e.d.

Now, for any left A^e-module M, let $\text{Der}_R(A,M)$ denote the R-module of R-linear maps $d: A \to M$ such that $d(ab) = (da)b + a(db)$ for all a, b in A. Such d are called <u>derivations</u>. 4.8(b) says that δ is a derivation from A to J. One sees easily (exercise) that if d is in $\text{Der}_R(A,M)$ and f is in $\text{Hom}_{A^e}(M,N)$ then fd is in $\text{Der}_R(A,N)$. Fix x in M and let f_x be the A^e-linear map from J to M given by $f_x(a \otimes b^o) = (a \otimes b^o)x = axb$. Let d_x denote the element $f_x\delta$ of $\text{Der}_R(A,M)$. The map $x \mapsto d_x$ from M to $\text{Der}_R(A,M)$ so defined is R-linear; derivations in the image are called <u>inner</u> derivations. Explicitly, $d_x a = (\delta a)x = ax - xa$ (for x in M, a in A).

4.10 Proposition. $f \mapsto f\delta$ <u>is an isomorphism</u>, $\text{Hom}_{A^e}(J,M) = \text{Der}_R(A,M)$. <u>The inner derivations correspond precisely to the A^e-linear maps from J to M which extend to all of A^e</u>.

<u>Proof</u>. The second part is clear. The map is injective because of 4.8(a): if f is zero on the image of δ, f is zero. For surjectivity, given d in $\text{Der}_R(A,M)$, define $f_d: A^e \to M$ by $f_d(a \otimes b^o) = -a(db)$. An easy computation shows that $f_d\delta = d$ (show first that $d(1) = 0$). So we are done if the restriction of f_d to J is A^e-linear. But if $x = \Sigma a_i \otimes b_i^o$ is in J, the relation $\Sigma a_i b_i = 0$ gives the next-to-last step of

$f_d((a \otimes b^o)x) = f_d \Sigma aa_i \otimes (b_i b)^o = -\Sigma aa_i d(b_i b) = -\Sigma aa_i((db_i)b + b_i(db)) = -\Sigma aa_i(db_i)b = (a \otimes b^o)f_d x$, q.e.d.

The preceeding results give us, for any left A^e-module M, a commutative diagram

$$0 \to \text{Hom}_{A^e}(A,M) \to \text{Hom}_{A^e}(A^e,M) \to \text{Hom}_{A^e}(J,M) \to 0$$
$$\downarrow \qquad \qquad \downarrow \qquad \qquad \downarrow$$
$$0 \to M^A \to M \to \text{Der}_R(A,M) \to 0$$

of R-modules. The verticals are isomorphisms, and both rows are exact if we omit the right-hand zeroes. The upshot is the following criterion for separability:

4.11 Proposition. *The following are equivalent for an R-algebra* A :

(1) A *is R-separable*.

(2) *Every derivation of* A *(to any left* A^e-*module) is inner*.

(3) δ *in* $\text{Der}_R(A,J)$ *is inner*.

Proof. A is R-separable iff $E(A)$ is split exact, in which case $\text{Hom}_{A^e}(E(A),M)$ will be (split) exact for any left A^e-module M. Then $M \to \text{Der}_R(A,M)$ is onto, i.e. every derivation (from A to M) is inner. The converse follows from 4.10 : under the isomorphism $\text{Der}_R(A,J) \simeq \text{Hom}_{A^e}(J,J)$, δ corresponds to the identity map 1_J on J . Hence δ is inner iff 1_J extends to an A^e-linear map from A^e to J , i.e. iff $E(A)$ splits.

Note what it means for δ to be inner: for some x in J , $\delta = d_x = f_x \delta$, or explicitly: $\delta a = f_x(\delta a) = (\delta a)x$ for all a in A.

The criterion given by 4.11 allows us to prove:

4.12 __Theorem__. Let R be a __local ring with maximal ideal__ m, A __an__ R-__algebra, finitely generated as__ R-__module. Then__ A __is__ R-__separable if and only if__ $\bar{A} = A/mA$ __is__ $\bar{R} = R/m$-__separable__.

(More is true: if A is an algebra, finitely generated as module, over __any__ commutative ring R, then A is R-separable iff A/mA is R/m-separable for every maximal ideal m of R. For an indication of how this is proved, see DEMEYER & INGRAHAM [1], Chapter II §7; for a simpler proof, under the stronger hypothesis that A is finitely presented as R-module, see BASS [1], Chapter III §3. We prove here only the local version, which is all we shall need.)

__Proof of 4.12__. The "only if" part follows from 2.3(c). For the converse note first that the canonical derivation $\bar{\delta} : \bar{A} \to \bar{J}$ is induced by the canonical derivation $\delta : A \to J$ (i.e., "$\bar{\delta} = \overline{\delta}$"). (Here $\bar{J} = R/m \otimes_R J$, $\bar{\delta} = R/m \otimes_R \delta$.) Since $\bar{\delta}$ is inner, we have $\bar{x}$ in $\bar{J}$ with $\bar{\delta}\bar{a} = (\bar{\delta}\bar{a})\bar{x}$ for all $\bar{a}$ in $\bar{A}$. Choose pre-images a in A for $\bar{a}$ and x in J for $\bar{x}$; the equation then reads $\delta a \equiv (\delta a)x \pmod{mJ}$. Because of 4.8(a), this yields $J = Jx + mJ$, hence by Nakayama's lemma (1.9), $J = Jx$. (Note that J is finitely generated, even though R might not be Noetherian, because A is finitely generated and the image of δ generates J as A^e-module.) Hence right multiplication by x is a surjective A^e-endomorphism of J. But it is then an __isomorphism__ by 1.14. Since it obviously extends to give an element of $\text{Hom}_{A^e}(A^e, J)$, we can use it to split $E(A)$; done.

4.13 __Application__. Let R be a ring in which 2 is a unit. Let a, b be units in R. The quaternion algebra $\left(\frac{a,b}{R}\right)$ was defined in 2.21 : $\left(\frac{a,b}{R}\right) = R \oplus Ri \oplus Rj \oplus Rk$ with $i^2 = a$, $j^2 = b$, $ij = k = -ji$. For R a field, we saw that

$\left(\frac{a,b}{R}\right) \simeq (R)_2$ if and only if $ax^2 + by^2 = 1$ has a solution (x,y) in R (2.21(c)). We shall show that this result holds more generally whenever R is local.

Let m be the maximal ideal of R, and suppose $ax^2 + by^2 = 1$. Then x or y, say y, is not in m. Put

$$A = \begin{pmatrix} 0 & a \\ 1 & 0 \end{pmatrix} \quad , \quad B = \frac{1}{y}\begin{pmatrix} 1 & ax \\ -x & -1 \end{pmatrix} .$$

Then $A^2 = a$, $B^2 = b$, and $AB = -BA$. The map from $\left(\frac{a,b}{R}\right)$ to $(R)_2$ sending 1, i, j, k to 1, A, B, AB respectively is therefore a homomorphism; apply 2.12(b) to conclude it is injective and count dimensions mod m to conclude it is surjective.

Conversely, let $h : \left(\frac{a,b}{R}\right) \to (R)_2$ be an R-algebra isomorphism, and let $A = h(i)$, $B = h(j)$. We can assume neither a nor b is a square in R, since for example if $a = c^2$ then the equation $ax^2 + by^2 = 1$ has the solution $(x,y) = (1/c,0)$. Thus $f(X) = X^2 - a$ is the minimal polynomial of A. We claim that $S = R[X]/(f(X))$ is a separable R-algebra. For this it suffices by 4.12 to see that S/mS is R/m-separable. But $S/mS = R/m \otimes S = R/m[X]/(X^2 - \bar{a})$, and $X^2 - \bar{a}$ is a separable polynomial over R/m because char(R/m) ≠ 2; hence S/mS is either $R/m \times R/m$ (if $\bar{a}$ is a square in R/m) or a separable field extension of R/m. In either case, S/mS is R/m-separable (by 4.5, 4.7 respectively).

Now let R[X] act on $V = R \oplus R$ by $Xv = Av$. This makes V into an S-module, which is projective by 2.4. In fact, V is a <u>free</u> S-module. To begin with, S is semi-local by 4.3(b), and S is connected: if $S = S_1 \times S_2$, each S_i is a projective, hence free, R-module of rank 1, hence $S = R \times R$. But then projecting the fact that a is a square in S onto one of the factors shows that a is a square in R, contrary to hypothesis. Since S is connected, all finitely generated projective S-modules have constant rank (if this is unfamiliar, refer ahead to 5.10(e)), hence because S is semilocal as well, all such modules are free (by 1.13(b)).

Thus V is a free S-module, clearly of rank one (because $[V : R] = [V : S][S : R]$). Hence there exists v in V such that $\{v, Av\}$ is a basis for V over R. Relative to this basis, the matrix of A is $\begin{pmatrix} 0 & a \\ 1 & 0 \end{pmatrix}$. By applying the appropriate inner automorphism in $(R)_2$, we may assume $A = \begin{pmatrix} c & a \\ 1 & 0 \end{pmatrix}$. Since $AB = -BA$, B must have the form $B = \begin{pmatrix} c & -ad \\ d & -c \end{pmatrix}$. Since $B^2 = b$, we have $ad^2 + b = c^2$.

Now if c is a unit, $(x,y) = (d/c, 1/c)$ is a solution. If c is a non-unit, the equation $ad^2 + b = c^2$ shows that d is a unit. Thus it suffices to show: if a, b are units in R with $a + b = z^2$, z in m, then $ax^2 + by^2 = 1$ for some x, y in R.

The equations

$$ax + (z - a)y = 1$$
$$x + (1 - z)y = 1$$

have a solution (x,y) in R because the determinant $2a - z(a + 1)$ is a unit in R (cf. 1.16(a)). Then

$$a(x - y) = 1 - yz$$
$$x + y = 1 + yz,$$

hence $a(x^2 - y^2) = 1 - y^2 z^2 = 1 - y^2(a + b)$, and $ax^2 + by^2 = 1$, as desired.

4.14 <u>Exercise</u>. Let A be an S-algebra, S a commutative R-algebra.

(a) Show that A is S-separable if it is R-separable.

(b) Show that S is R-separable if A is R-separable and faithfully projective as S-module (cf. 1.4) .

(c) Show that S = R[X] is never R-separable. Use this to show that the assertion " S is R-separable if A is" is false without some further hypothesis.

CHAPTER 5. COMPLETE LOCAL RINGS.

In this chapter we prove that $B(R) \to B(R/\mathfrak{m})$ is injective if R is a complete local ring with maximal ideal $\mathfrak{m}$, and discuss some related results and examples.

Let $\mathfrak{a}$ be a two-sided ideal of a ring A. For each n we have a commuting diagram

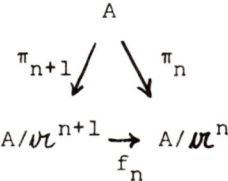

of natural projections. (By convention we include $n=0$, taking $\mathfrak{a}^0 = A$.) Hence there is an induced map $\nu: A \to \varprojlim(A/\mathfrak{a}^n)$. We say that A is $\mathfrak{a}$-<u>adically complete</u> if ν is an isomorphism. Note that injectivity of ν is equivalent to $\bigcap_n \mathfrak{a}^n = 0$, and surjectivity of ν is equivalent to either of the following:

<u>S1</u>. Given x_i in $A/\mathfrak{a}^i$ for each i, satisfying $f_i(x_{i+1}) = x_i$, there exists x in A with $\pi_i(x) = x_i$ for all i.

<u>S2</u>. Given $a_1, a_2, \ldots$ in A with $a_{i+1} - a_i$ in $\mathfrak{a}^i$ for all i, there exists a in A with $a - a_i$ in $\mathfrak{a}^i$ for all i.

In practice, <u>S2</u>. is usually easier to check.

Of course, <u>bijectivity</u> of ν is equivalent to existence of a <u>unique</u> element a as in <u>S2</u>. In this case one writes $a = \lim_{i \to \infty}(a_i)$. Note that $\mathfrak{a}$ is <u>closed</u> in the $\mathfrak{a}$-adic topology which we are trying not to mention: if $a = \lim_{i \to \infty}(a_i)$ and a_1 is in $\mathfrak{a}$ then the fact that $a - a_1$ is in $\mathfrak{a}^1 = \mathfrak{a}$ shows that a is in $\mathfrak{a}$. Thus:

5.1 **Lemma.** *Let* $a = \lim_{i \to \infty}(a_i)$. *The following are equivalent:*

(1) a is in $\mathcal{M}$.

(2) a_1 is in $\mathcal{M}$.

(3) a_i is in $\mathcal{M}$ for some $i \geq 1$.

(4) a_i is in $\mathcal{M}$ for all $i \geq 1$.

5.2 **Corollary.** *If A is $\mathcal{M}$-adically complete and z_i is in $\mathcal{M}^i$ for each i , any series $\sum_{j=0}^{\infty} d_j z_j$ (d_j in A) is a well-defined element of A ; it is in $\mathcal{M}$ if $d_0 = 0$.*

Proof. Let a_i denote the partial sum $\sum_{j=0}^{i-1} d_j z_j$; then $a = \sum_{j=0}^{\infty} d_j z_j$ denotes $\lim_{i \to \infty}(a_i)$. Since $a_1 = d_0 z_0$, $d_0 = 0$ implies $a_1 = 0$ implies (by 5.1) that a is in $\mathcal{M}$.

5.3 **Lemma.** *Let A be $\mathcal{M}$-adically complete. Given z in $\mathcal{M}$ there exists x in $\mathcal{M}$ such that*

(1) $x^2 - x = z$, *and*

(2) *For all y in A , $yz = zy$ if and only if $yx = xy$.*

Proof. (Jacobson). For $i = 1, 2, \ldots$ let $c_i = \frac{1}{2i-1}\binom{2i-1}{i}$. We leave to the reader the exercise of proving by induction that the c_i are all integers. Hence the c_i make sense as elements of A . Now, given z in $\mathcal{M}$, put $x = \sum_{i=1}^{\infty} c_i(-z)^i = -z + z^2 - 2z^3 + 5z^4 - \ldots$. 5.2 shows that x is a well-defined element of $\mathcal{M}$. Clearly if y is in A then $yz = zy$ implies $yx = xy$, and the converse is clear if (1) holds. Thus it remains to check that $x^2 - x = z$. This equation means $x_n^2 - x_n \equiv z \pmod{z^n}$, where $x_n = \sum_{i=1}^{n-1} c_i(-z)^i$, for all n . For example when $n = 4$ we have

$x_n^2 = (-z+z^2-2z^3)^2 \equiv z^2 - 2z^3 \pmod{z^4}$, and hence $x_4^2 - x_4 \equiv z \pmod{z^4}$. The reader can check (by induction, of course!) that the congruence holds for all n ;   q.e.d.

5.4 <u>Proposition</u>. <u>Let A be $\mathcal{M}$-adically complete and let $x \mapsto \bar{x}$ be the projection $A \to A/\mathcal{M}$. Given any idempotent f of $\bar{A}$, there is an idempotent e of A such that $\bar{e} = f$</u>. (Recall that "x is idempotent" means $x^2 = x$.)

<u>Proof</u>. Choose y in A with $\bar{y} = f$. Since $f^2 = f$ in $\bar{A}$ we have that $z = y^2 - y$ lies in $\mathcal{M}$. Note that $1 + 4z$ is invertible in A : $1 = (1+4z)(1-4z+16z^2-\ldots)$ and the series makes sense by 5.2, since A is $\mathcal{M}$-adically complete. Now apply 5.3 to the element $-z(1+4z)^{-1}$ of $\mathcal{M}$: there exists x in $\mathcal{M}$ such that

(*) $x^2 - x = -z(1 + 4z)^{-1}$.

Moreover, since $yz = y(y^2-y) = zy$, we have also

(**) $yx = xy$.

Finally, put $e = y(1-2x) + x$. Since $\bar{x} = 0$ we have $\bar{e} = \bar{y} = f$. Using (**) to compute e^2 , and (*) to simplify, we find $e^2 = e$; done.

5.5 <u>Lemma</u>. <u>Let R be a J-adically complete commutative ring (for some ideal J) and let A be an R-algebra which is finitely generated as an R-module. Then A is $\mathcal{M}$-adically complete</u>, where $\mathcal{M}$ = JA .

Proof. Since R is J-adically complete, $\cap J^n = 0$. Consequently $\cap \mathcal{M}^n = \cap J^n A = 0$, and $A \to \varprojlim(A/\mathcal{M}^n)$ is injective. To see that it is surjective, we check S2. Given $a_1, a_2, \ldots$ in A with $a_{i+1} - a_i$ in $\mathcal{M}^i$ for each i, put $a_0 = 0$ and $b_i = a_{i+1} - a_i$ for all $i \geq 0$. Then $\sum_{i=0}^{n-1} b_i = a_n$ for all $n \geq 1$. Let $x_1, \ldots, x_m$ generate A as R-module. Then since each b_i is in $\mathcal{M}^i$ we can write $b_i = \sum_{k=1}^{m} c_{ik} x_k$ with c_{ik} in J^i. Hence for each $n \geq 1$ we have $a_n = \sum_{i=0}^{n-1} b_i = \sum_{k=1}^{m} \alpha_{nk} x_k$ where $\alpha_{nk} = \sum_{i=0}^{n-1} c_{ik}$. Since $\alpha_{n+1,k} - \alpha_{nk} = c_{nk}$ is in J^n, J-adic completeness of R gives us a (unique) element α_k in R with $\alpha_k - \alpha_{nk}$ in J^n for all n, for each $k = 1, \ldots, m$. Now put $a = \sum_{k=1}^{m} \alpha_k x_k$. For each n, $a - a_n = \sum_k (\alpha_k - \alpha_{nk}) x_k$ is in $\mathcal{M}^n$, and we are done.

We adopt the following convention: if we say simply A <u>is complete</u>, without specifying any ideal, we mean A <u>is $\mathcal{M}$-adically complete where</u> $\mathcal{M} = \mathrm{rad}(A) =$ intersection of all maximal (left or right) ideals of A.

5.6 <u>Exercise</u>. Show that if $\mathcal{M}$ is nilpotent ($\mathcal{M}^N = 0$ for some N) then A is $\mathcal{M}$-adically complete. Show that if A is Artinian then $\mathrm{rad}(A)$ is nilpotent. Hence Artinian rings (and in particular fields, finite rings,...) are complete.

We can now prove:

5.7 <u>Theorem</u>. <u>If</u> R <u>is a complete local ring</u>, $B(R) \to B(R/\mathcal{M})$ <u>is injective</u>.

Proof. Let A be a central separable R-algebra such that $[\bar{A}] = 1$ in $B(\bar{R})$, where $\bar{R} = R/\mathcal{M}$, $\bar{A} = R/\mathcal{M} \otimes A = A/\mathcal{M}A$. We may assume $\bar{A} = (\bar{R})_n$. Let f be the matrix with 1 in the northwest corner, 0 elsewhere. Clearly $\text{End}_{\bar{R}}(\bar{A}f) \simeq \bar{A}$, because $\bar{A}f \simeq \bar{R}^n$. By 5.5, A is $\mathcal{M}A$-adically complete; hence by 5.4 there is an idempotent e in A with $\bar{e} = f$. Since $A = Ae \oplus A(1-e)$, Ae is a finitely generated projective, and hence a free, R-module, and $Ae \neq 0$ because $f \neq 0$. Hence if we show that the map $\eta : A \to \text{End}_R(Ae)$, $\eta(a) = $ left multiplication by a, is an isomorphism of R-algebras, we are done. It is certainly a homomorphism, and it is injective because Ae is a faithful left A-module by 2.12(a). To see that η is surjective it suffices, by Nakayama's lemma (1.9), to see that $\bar{\eta} : \bar{A} \to \overline{\text{End}_R(Ae)}$ is surjective. Now $\bar{\eta}$ is injective because $\bar{A}$ is simple, hence $\bar{\eta}$ is surjective if $\dim_{\bar{R}}(\bar{A}) = \dim_{\bar{R}}(\overline{\text{End}_R(Ae)})$. But $\overline{\text{End}_R(Ae)} \simeq \text{End}_{\bar{R}}(\bar{A}\bar{e}) = \text{End}_{\bar{R}}(\bar{A}f) \simeq \bar{A}$ (the first $\simeq$ is an easy exercise and the rest was done above); q.e.d.

5.8 Remarks and Examples.

(a) In fact the map in 5.7 is surjective as well; see theorem 6.5 of AUSLANDER & GOLDMAN [2]. F. DeMeyer has recently proved that $B(R) \to B(R/m)$ is surjective for certain other local rings R; see DEMEYER [4]. On the other hand there are easy examples to show that injectivity fails in general: see (e) and (f) below.

(b) The kind of proof we have given, by lifting idempotents, is unfashionable nowadays. It could be replaced by a Hensel's lemma argument, thereby proving 5.7 (and the surjectivity referred to in 5.8(a) as well) for Henselian local rings, of which complete local rings are merely the most conspicuous examples.* In fact the original proof of 5.7 in AZUMAYA [2], is Henselian; AUSLANDER & GOLDMAN [2] revert to the technique of lifting idempotents.

(c) We defined $\mathfrak{m}$-adic completeness, without defining $\mathfrak{m}$-adic topologies or completions with respect to them. The reader who wants more details should consult, e.g., NAGATA [1], §§ 16-17. We will, however, need some basic facts about the completion of a local ring (in its $\mathfrak{m}$-adic topology), which we sketch here.** Let R be a local ring and, for simplicity, assume that $\bigcap \mathfrak{m}^n = 0$. (This is automatic if R is Noetherian; see NAGATA [1], Theorem 4.2.) The $\mathfrak{m}$-adic topology on R is given by the following metric (exercise: verify the axioms): $d(x,x) = 0$ for all x; $d(x,y) = (1/2)^n$ if $x - y$ is in $\mathfrak{m}^n$ but not in $\mathfrak{m}^{n+1}$. (Replacing 1/2 by any ε such that $0 < \varepsilon < 1$ gives, of course, the same topology.) Since we have a metric topology we have the notion of Cauchy sequence, and the completion $\hat{R}$ of R, whose elements are equivalence classes of Cauchy sequences, is constructed in the usual way. Then $\hat{R}$ is a complete local ring containing R, whose maximal ideal is $\hat{\mathfrak{m}} = \mathfrak{m}\hat{R}$; $\hat{R}$ is Noetherian if R was; and the inclusion $R \to \hat{R}$ induces an isomorphism $R/\mathfrak{m} \to \hat{R}/\hat{\mathfrak{m}}$. For proofs, see NAGATA [1], §17, or chapter III of BOURBAKI [2]. (The fact that $R/\mathfrak{m} \to \hat{R}/\hat{\mathfrak{m}}$ is an isomorphism follows from Nagata's 17.9 with the trivial observation that fields are complete local rings.)

---

* For definition of Henselian local ring, and a proof that complete implies Henselian, see Nagata's "Local Rings", pp. 103-4. But note that what we are calling local ring, local ring satisfying $\bigcap \mathfrak{m}^n = 0$, and Noetherian local ring, are called by Nagata quasi-local ring, local ring which may not be Noetherian, and local ring, respectively; and that, in Nagata's terminology, a complete quasi-local ring is automatically a complete local ring which may not be Noetherian.

** A footnote due to Jacobson ("Structure of Rings", Revised Edition (1964), p. 256) is apropos here: "We should note...that our brief sketch...slurs over some of the difficulties of the method and may therefore be misleading. The reader should therefore consult the reverences for the discussion of this method."

(d) (Exercise) Let R, $\mathfrak{m}$ be a local ring such that $\bigcap \mathfrak{m}^n = 0$ and let $\hat{R}$, $\hat{\mathfrak{m}}$ be its completion. Conclude from the commutative diagram

$$\begin{array}{ccc} B(R) & \longrightarrow & B(\hat{R}) \\ \downarrow & & \downarrow \\ B(R/\mathfrak{m}) & \longrightarrow & B(\hat{R}/\hat{\mathfrak{m}}) \end{array}$$

that $B(R) \to B(\hat{R})$ and $B(R) \to B(R/\mathfrak{m})$ have the same kernel; in particular one is injective if and only if the other is. Examine the analogous remark for surjectivity (cf. (a)).

(e) Let R be the localization $\mathbb{Z}_{(p)}$ of $\mathbb{Z}$ at an odd prime p. We show that $B(R) \to B(R/\mathfrak{m})$ is not injective. Since $\mathfrak{m} = p\mathbb{Z}_{(p)}$ we have $B(R/\mathfrak{m}) = B(\mathbb{Z}/p\mathbb{Z}) = 0$ by 3.7; hence it suffices to see that $B(R) \neq 0$. But 2 is invertible in R (p was odd), and therefore the quaternion algebra $A = \left(\frac{-1,-1}{R}\right)$ is central separable (2.20); [A] is non-trivial in $B(R)$ because $[\mathbb{R} \otimes_R A]$ is nontrivial in $B(\mathbb{R})$.

(f) (Exercise; O. Goldman) Let R be the localization of $\mathbb{R}[X]$ at the prime ideal $(X^2 + 1)$, and let $\mathfrak{m}$ be the maximal ideal of R. Show that $B(R) \to B(R/\mathfrak{m})$ is not injective.

(g) Keep the notation of (e): R is the localization $\mathbb{Z}_{(p)}$ of $\mathbb{Z}$ at a prime p. Let $\hat{R} = \mathbb{Z}_p$ denote the completion of $R = \mathbb{Z}_{(p)}$ (thus $\mathbb{Z}_p$ is the ring of p-adic integers). Since the residue class field of $\hat{R}$ is $\mathbb{Z}/p\mathbb{Z}$, 3.7 and 5.7 conspire to show that $B(\hat{R}) = 0$. Since we just saw that $B(R) \neq 0$ (at least for odd p), this shows that completion can kill elements of the Brauer group. (Of course, we knew this before observing that $B(\hat{R}) = 0$, by (d) and (e).)

If there were sufficient justice in the world, the fact (just noted) that $B(\mathbb{Z}_p) = 0$ for all p should imply that $B(\mathbb{Z}) = 0$. (Here "justice" means something like "Hasse-Minkowski".) In fact this is (cum grano salis) true, as we shall see in proving 6.36 — but we will need also the deep theorem 3.12 about $B(\mathbb{Q})$. This raises the following Question: if $B(\hat{R}_p)$ is trivial for all prime ideals p of R, is $B(R)$ trivial? More generally, what can be said about the kernel (and image) of $B(R) \to \prod B(\hat{R}_p)$? For results about the kernel of $B(R) \to \prod B(R_p)$ (without $\hat{}$) see B. AUSLANDER [2], CHILDS [4] and [5], and OJANGUREN [1]; cf. also Cor. 1.10 of GROTHENDIECK [2]. In the next chapter we will see that if R is a regular domain with quotient field K, then $B(R) \to B(K)$ is injective, from which it follows easily that $B(R) \to \prod_p B(R_p)$ is injective.

5.9 Corollary. Let R be a finite ring. Then $B(R) = 0$.

Proof. If R is local it is complete by 5.6 and we are done by 5.7 and 3.7. On the other hand 5.10 (c) and (i) below show that we can assume R is connected. Hence we are reduced to proving that if R is finite and connected it is local. The following shows that this holds even with "finite" replaced by "Artinian": Let $S = R/\mathrm{rad}(R)$; we show S is a field. Since $\mathrm{rad}(S) = 0$ and S is Artinian, we have that S is semisimple (see e.g. BASS [2], chapter III). So, by the structure theory for (commutative!) semisimple rings, S is a product of fields. But S is connected because R is : R is complete (this was noted in 5.6), and any non-trivial idempotent of S would therefore lift, by 5.4, to R. Thus S is a field, and we are done.

The following exercise summarizes some standard material. Parts were used in proving 5.9, and others will be needed in chapter 12.

5.10 **Exercise**. Call R **connected** if it has no non-trivial idempotents: $e^2 = e$ in R implies $e = 0$ or $e = 1$.

(a) If R is local, or a domain, then R is connected. Find connected R which are neither local nor a domain.

(b) Show that $R \simeq Re \times R(1 - e)$ if e is a nontrivial idempotent in R. Conversely, if $R = R_1 \times R_2$ where neither R_i is the zero ring, R is not connected.

(c) R is a finite direct product of connected rings if, and only if, R has only finitely many idempotents.

(d) If R is Noetherian, it satisfies the equivalent properties of part (c). (Hint: for classical reasons, R has only finitely many minimal prime ideals. But whenever $R \simeq R_1 \times \ldots \times R_N$ and p_i is a minimal prime of R_i, $R_1 \times \ldots \times R_{i-1} \times p_i \times R_{i+1} \times \ldots \times R_N$ is a minimal prime in R.)

(e) If R is connected, every finitely generated projective R-module has constant rank. (The conceptual argument goes as follows: given a module P, its rank is a **continuous** function $\text{Spec}(R) \to \mathbb{Z}$, where $\text{Spec}(R)$ (resp. $\mathbb{Z}$) has the Zariski (resp. the discrete) topology. Hence rank is always constant on connected components of $\text{Spec}(R)$. But $\text{Spec}(R)$ is connected if (and only if) R is.)

(f) If $R = R_1 \times R_2$ and M is an R-module, let $M_1 = M/R_2 M = R_1 \otimes_R M$ and $M_2 = M/R_1 M = R_2 \otimes_R M$. Show that $M \mapsto (M_1, M_2)$ is a category equivalence from ((R-modules)) to ((R_1-modules)) $\times$ ((R_2-modules)). Similarly, $\text{Spec}(R)$ is the disjoint union of $\text{Spec}(R_1)$ and $\text{Spec}(R_2)$.

(g) Keeping the notation of (f), M is a finitely generated projective R-module if and only if M_i is a finitely generated projective R_i-module $(i = 1, 2)$. Any prime ideal p of R "is" a prime ideal p_i of R_i for $i = 1$ or 2 but not both (cf. part (f)), and the rank of M at p equals the appropriate rank of M_i at p_i.

(h) Given a finitely generated projective R-module P, let $\mathbb{P}_n$ be the set of prime ideals of R at which the rank of P is n (for each $n \geq 0$). Show that $\{n | \mathbb{P}_n \neq \emptyset\}$ is finite. (Outline: $\mathbb{P}_n$ is the inverse image, under the continuous map described in part (e), of the open set $\{n\}$. The $\mathbb{P}_n$ therefore form an open cover of $\text{Spec}(R)$. But $\text{Spec}(R)$ is quasi-compact.) Hence P yields a finite decomposition of $\text{Spec}(R)$, which gives a decomposition $R = R_1 \times \ldots \times R_m$ of R with the property that, writing $P = P_1 \times \ldots \times P_m$ (cf. part (f)), P_i is a finitely generated projective R_i-module of constant rank.

(i) Let $R = R_1 \times R_2$. Starting from the first part of (f), show that $A \mapsto (A_1, A_2)$ is a category equivalence from ((R-algebras)) to ((R_1-algebras)) $\times$ ((R_2-algebras)), and that A is central separable over R iff A_i is central separable over R_i $(i = 1, 2)$. Conclude that $B(R) \simeq B(R_1) \oplus B(R_2)$.

CHAPTER 6. THE BRAUER GROUP OF SOME SPECIAL DOMAINS.

The main results of this chapter are theorems 6.19 and 6.33 . The first of these says that $B(R) \to B(K)$ is injective if R is a regular domain with quotient field K . The second says that if, in addition, R has global dimension ≤ 2 , then $B(R)$ is the intersection in $B(K)$ of the $B(R_p)$, p ranging over all height 1 primes of R . (The notions of global dimension and regularity are discussed in 6.20 .) These two results are applied to compute $B(R)$ where R is the ring of integers in an algebraic number field (6.36) .

Let K be a field, Σ a central simple K-algebra, assumed finite dimensional over K . We know that Σ admits a splitting field L which is a finite Galois extension of K (3.15) . Choose such an L , and an L-algebra isomorphism $h : L \otimes_K \Sigma \to (L)_n$, and define the <u>reduced trace</u> of an element a of Σ by $tr(a) = trace(h(1 \otimes a))$.

The map tr is independent of the choice of h . For if g is another isomorphism from $L \otimes \Sigma$ to $(L)_n$ then $h^{-1}g$ must be an inner automorphism of $L \otimes \Sigma$ (3.6) ; $h(1 \otimes a)$, $g(1 \otimes a)$ are then similar matrices and therefore have the same trace.

Actually, $tr(\Sigma) \subseteq K$. For if σ is in the Galois group of L/K , then, by independence from h , $tr(a) = trace(\sigma_n h(1 \otimes a)) = \sigma(trace\, h(1 \otimes a)) = \sigma tr(a)$, where σ_n is the automorphism of $(L)_n$ induced entrywise by σ . Finally, we note that tr is independent of the choice of L : the case $L \subseteq L'$ follows from diagram chasing, and one then uses the fact that any two Galois extensions are contained in some common Galois extension.

6.1 __Exercise.__ (a) Let A be a central simple K-algebra. Choose a Galois splitting field L for A and an isomorphism $h : L \otimes A \to (L)_n$. Define $N_{red}(a) = \det(h(1 \otimes a))$ for a in A. Show that the reduced norm of x thus defined is independent of h, lies in K, and is independent of the splitting field L chosen. Show that N_{red} is multiplicative.

(b) Let $a_1,\ldots,a_m$ be a basis of A over K. For an element $a = \Sigma x_i a_i$ in A, $N_{red}(a) = f(x_1,\ldots,x_m)$ ($m = n^2$) where f is a homogeneous polynomial (a form) in $K[X_1,\ldots,X_m]$ of degree n.

6.2 __Lemma.__ __Let__ Σ __be a central simple__ K-__algebra. Define a map__ $\psi_\Sigma : \Sigma \to \Sigma^*$ ($\Sigma^* = \text{Hom}_K(\Sigma,K)$) by $\psi(x)(y) = tr(xy)$. __Then__ ψ __is an isomorphism.__

__Proof.__ Let $L \otimes \Sigma \simeq (L)_n$, with L/K Galois. It suffices to show $1 \otimes \psi$ (and hence ψ) is monic. But $1 \otimes \psi = \psi_{L \otimes \Sigma}$ when the obvious identification is made, i.e., the diagram below commutes:

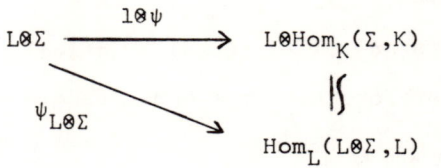

Hence we may assume Σ is a matrix ring over K. Let $0 \neq x$ be in Σ. By column operations one can transform x to a matrix of trace 1, i.e., $\text{trace}(xy) = 1$ for some y. Then $\psi(x)(y) \neq 0$, hence $\psi(x) \neq 0$.

6.3 __Proposition.__ __Let__ K __be a field containing__ R __as a subring. Let__ Σ __be a central simple__ K-__algebra and let__ x __be an element of__ Σ __which is integral over__ R. __Then__ $tr(x)$ (__the__ reduced trace of x) is integral over R .

Proof. Let $h : L \otimes_K \Sigma \to (L)_n$ be an L-algebra isomorphism; as discussed above $tr(x) = $ trace $h(1 \otimes x)$. The matrix $\alpha = h(1 \otimes x)$ is integral over R . View α as an endomorphism of $V = L^n$. Then V is a direct sum of α-cyclic subspaces, i.e., $V = V_1 \oplus \ldots \oplus V_k$ where $\alpha(V_i) \subseteq V_i$ and each V_i admits a basis of the form $v, \alpha v, \ldots, \alpha^s v$. In particular, the minimal and characteristic polynomials of the restriction α_i of α to V_i are the same. The trace of α is the sum of the traces of α_i , hence it suffices to show that $tr(\alpha_i)$ is integral over R . Fix i and let $\beta = \alpha_i$, an endomorphism of V_i . Clearly β is integral over R , satisfying the same equation as α does, say $f(\beta) = 0$ with $f(X)$ in $R[X]$. Let $m(X)$ be the minimal (and characteristic) polynomial of β in $L[X]$. Then $f(X) = m(X)g(X)$ for some monic $g(X)$ in $L[X]$. Let E be a splitting field of $m(X)$ over L and let $m(X) = \Pi(X-a_i)$, with a_i in E . Each a_i is integral over R , since $f(a_i) = 0$. Hence trace$(\beta) = \Sigma a_i$ is integral over R .

For the rest of this chapter let R be a Noetherian, integrally closed domain, and K its quotient field.

6.4 Definition. Suppose Σ is a finite dimensional K-algebra. An R-order in Σ is an R-algebra A contained in Σ , containing a K-basis of Σ , and such that every element of A is integral over R .

If A is an R-algebra which is contained in Σ , and is finitely generated as an R-module, then every element of A is integral over R (KAPLANSKY [2], p. 9) . The converse is true when

Σ is central simple:

6.5 Theorem. _Let_ Σ _be a central simple_ K-_algebra and_ A _an_ R-_order in_ Σ. _Then_ A _is contained in a maximal_ R-_order, and_ A _is finitely generated as_ R-_module._

Proof. The existence of a maximal R-order containing A follows by Zorn's Lemma. Let $u_1,\ldots,u_n$ be a K-basis of Σ contained in A, and let $L = Ru_1 \oplus \ldots \oplus Ru_n$. Let $\{f_1,\ldots,f_n\}$ be a basis of $\text{Hom}_K(\Sigma,K)$ dual to $\{u_1,\ldots,u_n\}$, i.e., $f_i(u_j) = \delta_{ij}$. Let $\psi_\Sigma(v_i) = f_i$ (using 6.2), and $L_1 = Rv_1 \oplus \ldots \oplus Rv_n$.

By choice of v_i, $\text{tr}(v_i x) = f_i(x)$ for all x in Σ. Since $f_i(L) = R$, we have $\text{tr}(v_i L) = R$. In fact

$$L_1 = \{x \text{ in } \Sigma \mid \text{tr}(xL) \subseteq R\},$$

for, writing $x = \Sigma a_i v_i$, a_i in K we see that $\text{tr}(xu_j) = \text{tr}(\Sigma a_i v_i u_j) = \Sigma a_i f_i(u_j) = a_j$.

By 6.3, $\text{tr}(A) \subseteq R$ (R is integrally closed). Then $\text{tr}(AL) \subseteq \text{tr}(A) \subseteq R$, hence $A \subseteq L_1$, and, R being Noetherian, this implies that A is finitely generated.

We shall establish in 6.14 and 6.15 below that R-orders exist in any finite-dimensional Σ.

6.6 Example. Let V be a finite dimensional vector space over K. An R-submodule E of V is called an R-_lattice_ if it is finitely generated and $KE = V$, i.e., E contains a K-basis of V (see 6.8 below). Let $v_1,\ldots,v_n$ be such a basis.

By writing each of the generators of E in terms of the v_i we see that $\oplus Rv_i \subseteq E \subseteq \oplus R(1/a)v_i$ for a suitable element a in R. Let $w_i = (1/a)v_i$, $i = 1,\ldots,n$.

In $\Sigma = \mathrm{End}_K(V)$ let e_{ij} satisfy $e_{ij}(w_k) = \delta_{jk}v_i$. Then $e_{ij}(E) \subseteq E$, hence $\mathrm{End}_R(E)$ contains a basis $\{e_{ij}\}$ of Σ over K. By exercise 6.7 below, every element of $\mathrm{End}_R(E)$ is integral over R. <u>Thus if</u> E <u>is an</u> R-<u>lattice in</u> V, $\mathrm{End}_R(E)$ <u>is an</u> R-<u>order in</u> $\mathrm{End}_K(V)$.

6.7 <u>Exercise</u>. (a) If R is any commutative Noetherian ring and M, N are finitely generated R-modules, then $\mathrm{Hom}_R(M,N)$ is a finitely generated R-module.
(b) If R is any commutative ring and M is a finitely generated R-module, then every element of $\mathrm{End}_R(M)$ is integral over R. (Use the Cayley-Hamilton theorem (4.1).)

6.8 <u>Exercise</u>. Let V be a finite dimensional vector space over K. Show that V contains an R-lattice E.

The construction carried out in 6.5 can be imitated in a slightly different setting. Let E be an R-lattice in a finite dimensional K-vector space V (see 6.6 for the definition). Given f in $E^* = \mathrm{Hom}_R(E,R)$, f extends uniquely to an element of $V^* = \mathrm{Hom}_K(V,K)$; this extension is also called f. Now $KE^* = V^*$, and E^* is a finitely generated R-submodule of V^* (6.7). We may characterize E^* in V^* as:

(6.9) $\quad E^* = \{f \text{ in } V^* \mid f(E) \subseteq R\}$.

Since E^* is now an R-lattice in V^*, we can repeat the process; hence E^{**} is an R-lattice in V^{**}. The natural isomorphism $\eta : V \to V^{**}$ induces a monomorphism from E to E^{**}, which we treat as an inclusion. Indentifying E^{**} with its inverse image $\eta^{-1}(E^{**})$ we may write $E \subseteq E^{**} \subseteq V$, with E^{**} an R-lattice in V characterized as:

(6.10) $\quad E^{**} = \{x \text{ in } V \mid f(x) \text{ is in } R \text{ for all } f \text{ in } E^*\}$.

6.11 <u>Proposition</u>. <u>Let A be an R-order in a central simple K-algebra Σ</u>. <u>Then A^{**} is an R-order in Σ, and the natural map $A \to A^{**}$ is an inclusion.</u>

<u>Proof</u>. After the discussion above (taking $V = \Sigma$) all that remains to be proved is that A^{**} is a ring. The Σ-Σ biaction on Σ makes Σ^* into a Σ-Σ bimodule: $(afb)x = f(bxa)$. A^* is clearly an A-A submodule of Σ^* since A is a ring.

Let x, y be in A^{**}. For xy to be in A^{**} it is sufficient, by 6.10, that $f(xy)$ be in R for all f in A^*. But $f(xy) = (fx)(y)$ by the nature of the Σ-Σ biaction on Σ^* mentioned above. If fx were in A^*, we would be done by 6.10. But for z in A, $(fx)(z) = f(xz) = (zf)(x)$ which is in R by 6.10 since x is in A^{**}.

6.12 <u>Corollary</u>. <u>Let A be a maximal R-order in a central-simple K-algebra Σ. Then A is a reflexive R-module</u>, i.e., <u>the natural map $A \to A^{**}$ is an isomorphism.</u>

6.13 <u>Exercise</u>. The following considerations pertaining to reflexivity of modules will be useful later. Let M be any R-module. Write $\eta : M \to M^{**}$ for the natural map, defined by $\eta(m)(f) = f(m)$ for f in M^*.

(a) Show that $\eta^* : M^{***} \to M^*$ is a splitting for the natural map $M^* \to M^{***}$, i.e., $(\eta_M)^* \eta_{M^*} = 1_{M^*}$.

Now let R be an integral domain with quotient field K.

(b) M^* is torsion-free.

(c) Let V be a finite dimensional vector space over K, $M \subseteq N$ R-lattices in V. Then $N^* \subseteq M^*$ in V^* (see 6.9). Using this, (a) above, and the discussion preceding 6.10, conclude that $M^* = M^{***}$; i.e., <u>if</u> M <u>is an</u> R-<u>lattice</u>, M^* <u>is always reflexive</u>.

6.14 <u>Example</u>. We shall show that any finite dimensional K-algebra Σ contains an R-order which is also an R-free R-lattice. Let $v_1 = 1$, $v_2,\ldots,v_n$ be a K-basis for Σ. Then $v_i v_j = \sum_k a_{ijk} v_k$, with a_{ijk} in K and $a_{1jk} = \delta_{jk} = a_{j1k}$. Let s be in R, with $sa_{ijk} = b_{ijk}$ in R for all i, j, k. Let $w_1 = 1$, $w_i = sv_i$ for $i = 2,\ldots,n$. It is easily seen that $A = \Sigma R w_i = \oplus R w_i$ is a free R-lattice as well as a ring containing R. To see that it is an R-order it suffices to show any element x of A is integral over R. But multiplication by x is an R-endomorphism of A, hence integrality of x follows by 6.7(b).

6.15 <u>Example</u>. The existence of R-orders can also be established by exhibiting certain special ones. Let E be an R-lattice in the finite dimensional K-algebra Σ (see 6.8). Let

$$O_\ell(E) = \{x \text{ in } \Sigma \mid xE \subseteq E\},$$
$$O_r(E) = \{x \text{ in } \Sigma \mid Ex \subseteq E\}.$$

These are called respectively the <u>left</u> and <u>right orders</u> of E in Σ. Each of $O_\ell(E)$, $O_r(E)$ is an R-algebra containing a basis of Σ over K. Every element of $O_\ell(E)$ is an endomorphism of a finitely generated R-module, hence integral over R, by 6.7(b).

6.16 <u>Exercise</u>. Let A be an R-order in the central simple K-algebra Σ. Then A is also an R-lattice in Σ, hence is sandwiched between free R-modules F_1 and F_2 of rank equal to $\dim_K \Sigma$ (see 6.6). Hence the multiplication map $K \otimes_R A \to \Sigma$ is an isomorphism.

6.17 <u>Remark</u>. It is an easy consequence of the above discussion, but worth noting explicitly, that if B is a subring of a finite dimensional K-algebra Σ, containing R and a basis of Σ over K, and finitely generated as an R-module, then B is an R-order in Σ. Hence B is contained in a maximal R-order.

The relevance of orders to the Brauer group is indicated by the following result.

6.18 <u>Proposition</u>. <u>Let A be a central separable R-algebra. Then A is a maximal order in</u> $K \otimes_R A$.

<u>Proof</u>. A is clearly an R-order in $\Sigma = K \otimes_R A$ ($A \subseteq K \otimes_R A$ since, being R-projective, A is also R-flat (1.17)). Suppose B is an R-order containing A. By 2.13, the multiplication map $A \otimes B^A \to B$ is an isomorphism. Then B^A is a finitely generated R-module: take $f : A \to R$ with $f(1) = 1$ (1.4); f induces an epimorphism $A \otimes B^A \to B^A$ from B to B^A, and B is a finitely generated R-module (6.5). Now $B^A = B^{KA} = B^\Sigma \subseteq K$, and B^A, being finitely generated over R, is integral over R. But R is integrally closed, hence $B^A = R$

and hence $A = B$.

The next theorem is the focus for the above discussion. We shall first state it and then explain the terminology involved and some prerequisites for its proof.

6.19 Theorem. *Let R be a regular domain. Then the map $B(R) \to B(K)$ is a monomorphism.*

6.20 Remarks. A ring R is said to have global dimension $\leq n$ (written gl. dim $R \leq n$) if every R-module M admits a projective resolution of length $\leq n$, i.e., there exists an exact sequence

$$0 \to P_n \to P_{n-1} \to \ldots \to P_1 \to P_0 \to M \to 0$$

with P_i being R-projective. A commutative local ring is said to be regular if it is Noetherian and has finite global dimension. For R commutative Noetherian, gl. dim $R = \sup_p(\text{gl. dim } R_p)$ where p ranges over all the maximal ideals of R (NORTHCOTT [2], Theorem 10, p. 187) ; in fact p may range over all prime ideals, since if $p \subseteq m$, gl. dim $R_p \leq$ gl. dim R_m. We call R regular if every localization R_p is regular, i.e., iff every R_p is a regular local ring.

There are two striking properties of regular rings which we shall use without proof:

6.21 Theorem. *A regular local ring is factorial* (i.e., a UFD) (KAPLANSKY [2], Theorem 184, p. 135 or SERRE [4], Corollaire 4,

p. IV - 39).

6.22 **Theorem.** Let R be a regular domain and M a finitely generated reflexive R-module. Suppose $\text{End}_R(M)$ is projective. Then M is projective.

(Assuming 6.21, we shall give a relatively self-contained proof of 6.22 in the Appendix to Chapter 11. The crucial material needed is the existence of a nice splitting ring for the R-algebra $\text{End}_R(M)$ (which will be shown to be central separable over R).)

6.23 **Remark.** A regular domain R is Noetherian by definition, and integrally closed because every R_p is a UFD, hence integrally closed (KAPLANSKY [2], Theorem 54, p. 35; Theorem 50, p. 33). Therefore the results on orders previously obtained apply to such rings.

Proof of 6.19. Let A be a central separable R-algebra and let $\Sigma = K \otimes_R A$. Suppose $\Sigma \sim 1$ in $B(K)$, i.e., $\Sigma = \text{End}_K(V)$, $\dim_K V = n$. We claim that $A = \text{End}_R(E)$, where E is an R-lattice in V. Let $v_1, \ldots, v_n$ be a K-basis of V and let $F = \oplus Rv_i$. Then (identifying A with its image in $\text{End}_K(V)$), AF is a finitely generated R-module containing a basis of V, hence is an R-lattice. If $E = AF$ then there are natural inclusions $A \subseteq \text{End}_R(E) \subseteq \text{End}_K(V)$. Because A is a maximal R-order in Σ (6.18), we have that $A = \text{End}_R(E)$.

We remark that if R is a Dedekind domain, the proof is now complete. For E is a finitely generated torsion free R-module, and is therefore faithfully projective (CARTAN & EILENBERG [1], Proposition 4.1, p. 133).

We wish to show that $A = \text{End}_R(M)$, with M a finitely generated reflexive R-module. The natural isomorphism $V \simeq V^{**}$ induces an isomorphism $\text{End}_K(V) \simeq \text{End}_K(V^{**})$, sending f to f^{**}.

By restriction, $\operatorname{End}_R(E) \simeq \operatorname{End}_R(E)^{**} = \{f^{**} \mid f \text{ in } \operatorname{End}_R(E)\}$. But clearly $\operatorname{End}_R(E)^{**} \subseteq \operatorname{End}_R(E^{**})$. Now $M = E^{**}$ is a reflexive R-lattice in V^{**} (6.8 ff, 6.13 (c)), hence $\operatorname{End}_R(E^{**})$ is an R-order (6.6). Because $A = \operatorname{End}_R(E)$ was a maximal R-order in Σ, we conclude that $A = \operatorname{End}_R(M)$, with M a finitely generated reflexive (hence faithful by 6.13 (b)) R-module. By 6.22, M is projective, hence A is the trivial element in B(R).

6.24 <u>Example</u>. We shall show that $B(R) \to B(K)$ can fail to be monic if R is a UFD. Let $\mathbb{R}$ be the field of real numbers and $R = \mathbb{R}[x,y,z]/(x^2+y^2+z^2)$. Let $Q = \left(\frac{-1,-1}{R}\right)$ be the usual algebra of quaternions over R (see 2.20, 2.21). Q is a central separable R-algebra, whose image under the map $B(R) \to B(\mathbb{R})$ is the algebra of quaternions over $\mathbb{R}$, hence nontrivial (2.21 (d)). Therefore $[Q] \neq 1$ in $B(R)$.

In K, the quotient field of R, we have $-1 = (x/z)^2 + (y/z)^2$, hence $K \otimes_R Q \simeq (K)_2$, and the map $B(R) \to B(K)$ is therefore not monic. The ring R is a Noetherian integrally closed domain, and is in fact even a UFD but we shall not prove this here. R can even be replaced by its localization at $m = (x,y,z)$, hence $B(R) \to B(K)$ can fail to be monic even if R is local.

Our next goal will be to describe B(R) in terms of the groups $B(R_p)$ as p ranges over primes of height 1 (minimal nonzero primes). When p is of height 1, R_p is a <u>discrete valuation ring</u>, i.e., a Noetherian integrally closed local domain with a unique nonzero prime ideal, which must then be principal (KAPLANSKY [2], p. 67, or SERRE [2], Ch. I §2). R will continue to be a Noetherian integrally closed domain with quotient field K; later we will assume that R is more special, viz. a regular domain of global dimension (or Krull dimension) ≤ 2 .

6.25 <u>Definition</u>. Let A, B be R-orders in the finite dimensional K-algebra Σ . Define the <u>conductor of</u> A <u>to</u> B <u>in</u> Σ by

$$\underline{c} = \{x \text{ in } \Sigma \mid xA \subseteq B\} .$$

We remark that $\underline{c}$ is a left ideal of B , as well as a right A-module. In fact, $\underline{c}$ contains every right A-module contained in B . Moreover $\underline{c} \cap R \neq 0$: for if $\{v_i\}$, $\{w_i\}$ are bases of Σ in A, B respectively, then $A \subseteq KB$ implies that $v_i = \sum_j \frac{r_{ij}}{r_i} w_j$, hence $(\Pi r_i)A \subseteq B$ for suitable $r_i \neq 0$ in R .

6.26 <u>Proposition</u>. <u>Let</u> M <u>be a multiplicative set in</u> R . <u>Suppose</u> A <u>is a (maximal)</u> R-<u>order in the central simple</u> K-<u>algebra</u> Σ . <u>Then</u> A_M <u>is a (maximal)</u> R_M-<u>order in</u> Σ .

<u>Proof</u>. Only the statement about maximality isn't obvious. Suppose B is an R_M-order in Σ containing A_M . Let $I = \{a \text{ in } A \mid aB \subseteq A_M\}$. I is a two-sided ideal in A , and is nonzero by the last remark in 6.25 . Hence $KI = \Sigma$ since Σ is simple, and I is therefore an R-lattice (it is finitely generated because R is Noetherian). Then $\mathcal{O}_r(I) = \{x \text{ in } \Sigma \mid Ix \subseteq I\}$ is an R-order in Σ (6.15) containing A , hence is A by maximality of the latter.

Now choose b in B . Since $Ib \subseteq A_M$ and I is a finitely generated R-module, there exists m in M with $Imb \subseteq A$. It follows from the definition of I that in fact $Imb \subseteq I$. Hence mb is in $\mathcal{O}_r(I) = A$, i.e., $B \subseteq A_M$.

6.27 __Corollary__. __Let__ A __be an__ R-__order in the central simple__ K-__algebra__ Σ . A __is maximal if and only if__ A_m __is maximal for all maximal ideals__ m __of__ R .

__Proof__. If A ⊆ B then $(B/A)_m = 0$ for all m implies A = B .

6.28 __Remarks__. For R a commutative ring, the __Krull dimension__ of R is the supremum of all n for which there exists a chain of prime ideals $p_0 \subsetneq p_1 \subsetneq \ldots \subsetneq p_n$. We shall use without proof the fact that for R a regular local ring, the Krull and global dimensions of R are equal (SERRE [4], Corollaire 2, p. IV-39) . Since each of the dimensions involved for a ring R is the supremum of the corresponding dimensions over all R_p (6.20) , we shall refer to the __dimension__ of R as the common value of the Krull and global dimensions, for R Noetherian of finite global dimension.

We shall now state a series of results which will be needed for the proof of 6.33, the second main theorem of this chapter. The proofs will be given after the proof of 6.33 .

6.29 __Proposition__. __Let__ R __be a Noetherian domain of global dimension__ ≤ 2 . __Then any finitely generated reflexive__ R-__module is projective__. (Proof follows 6.33 .)

6.30 __Proposition__. __Let__ R __be a commutative Noetherian ring__. __Let__ A __be an__ R-__algebra with center__ R __which is a faithfully projective__ R-__module__. __Suppose__ A_p __is a central separable__ R_p-__algebra for all primes__ p __of height__ 1 __in__ R . __Then__ A __is a central separable__ R-__algebra__. (Proof follows 6.33 .)

6.31 __Proposition__. _Let_ A _be a maximal_ R-_order in the division ring_ D. _Then the matrix algebra_ $(A)_n$ _is a maximal_ R-_order in_ $(D)_n$. (Proof follows 6.33.)

6.32 __Proposition__. _Let_ R _be a discrete valuation ring_, A _a central separable_ R-_algebra and_ B _a maximal_ R-_order in_ $\Sigma = K \otimes_R A$. _Then_ $A \simeq B$. (Proof follows 6.33.)

Let $\mathbb{P}$ denote the set of primes of height 1 in R.

6.33 __Theorem__. _Let_ R _be a regular domain of dimension_ ≤ 2. _Then_ $B(R) = \bigcap_{p \text{ in } \mathbb{P}} B(R_p)$ (_as a subset of_ $B(K)$).

__Proof__. We note that the inclusions $B(R) \subseteq B(R_p) \subseteq B(K)$ implicit in the statement of the theorem are a consequence of 6.19. Then $B(R) \subseteq \bigcap B(R_p)$ is clear.

Let D be a central K-division algebra which is in $\bigcap B(R_p)$, i.e., for each p in $\mathbb{P}$, there exists a central separable R_p-algebra $A(p)$ such that $[K \otimes_{R_p} A(p)] = [D]$ in $B(K)$. Thus $K \otimes_{R_p} A(p) \simeq (D)_{n(p)}$ for some integer $n(p)$ (see remarks preceding 3.1). Let A be a maximal R-order in D. Since $K \otimes_R A \simeq KA$ (6.16), we have $D \simeq K \otimes_R A$. If A were central separable, we would be done; we shall prove this is so.

A is a reflexive R-module because it is a maximal R-order (6.12). Then 6.29 implies that A is R-projective.

The center of A is R, since any element of it is in the center K of D (A is an R-order in D) and also integral over R, and R is integrally closed. Knowing this, we are reduced by 6.30 to considering whether A_p is a central separable R_p-algebra.

We know that A_p is a maximal R_p-order in D, by 6.26. The matrix algebra $(A_p)_{n(p)}$ is then a maximal R_p-order in $(D)_{n(p)}$, by 6.31. Then $(A_p)_{n(p)} \simeq A(p)$ by 6.32, hence $(A_p)_{n(p)}$ is a central separable R_p-algebra. But $(A_p)_{n(p)} \simeq (R_p)_{n(p)} \otimes_{R_p} A_p$ and, by 2.15, A_p is therefore a central separable R_p-algebra. This completes the proof of 6.33, modulo 6.29-6.32.

Proof of 6.29. (L. Roberts). Let R have global dimension ≤ 2. Suppose M is any finitely generated torsion-free R-module. Take an exact sequence

$$0 \to L \to F \to M^* \to 0$$

with F free of finite rank. This is easily checked to yield an exact sequence $0 \to M^{**} \to F^* \to Q \to 0$, where Q is a submodule of L^*. Then Q is torsion-free, since L^* is torsion-free (6.13 (b)), and finitely generated since F^* is.

Now $Q \subseteq K \otimes_R Q$ since Q is torsion-free (see e.g., CARTAN & EILENBERG [1], Proposition 2.1, p. 130). Taking a (finite) basis of $K \otimes_R Q$ over K and expressing the generators of Q in terms of it, we can construct an embedding $Q \subseteq G$ with G a free R-module of finite rank. We have two exact sequences, with G, F^* free:

$$0 \to Q \to G \to G/Q \to 0$$
$$0 \to M^{**} \to F^* \to Q \to 0 \,.$$

Since R has global dimension ≤ 2, M^{**} is projective. Hence if M is reflexive, it is projective.

Proof of 6.30. It suffices to show that $\eta : A \otimes A^\circ \to \text{End}_R(A)$ is an isomorphism (2.14). For this it is enough to see that, for each maximal ideal m, the top arrow in the commutative diagram

$$\begin{array}{ccc} R_m \otimes (A \otimes A^\circ) & \xrightarrow{1 \otimes \eta} & R_m \otimes \text{End}_R(A) \\ \cong \downarrow & & \cong \downarrow \\ A_m \otimes A_m^\circ & \xrightarrow{\eta_m} & \text{End}_{R_m}(A_m) \end{array}$$

is an isomorphism. Hence we may change notation and assume R is local. Note that the hypothesis on A is preserved.

Once R is local, A^e and $\text{End}_R(A)$ are free of the same rank, viz. $(\text{rk}_R(A))^2$, and by 1.16 it suffices to check that $\det \eta$ is a unit. But if not, it is contained in a height one prime by Krull's Principal Ideal Theorem (KAPLANSKY [2], Theorem 142, p. 104), contradicting the hypothesis that $\det(\eta_p)$ is a unit at all such primes p.

It is clear that the proof just given works for a more general class of rings than Noetherian R.

Proof of 6.31. (This pretty proof is taken from REINER [1] (Theorem 9.1, p. 115).) Let $(A)_n \subseteq B \subseteq (D)_n$, with B an order. Let E be the set of all entries in the matrices occurring in B. It is clear that $A \subseteq E$, and we shall show that E is in fact an order in D. To do this we need to show that E is a ring and a finitely generated R-module.

Let x be in E; say x is the (k,ℓ)-entry of the matrix X in B. The matrices e_{1k} and $e_{\ell 1}$ lie in $(A)_n$ (e_{ij} has 1 in (i,j) spot, 0 elsewhere), hence in B; hence $e_{1k} X e_{\ell 1}$ is in B. But this is the diagonal matrix $\text{diag}(x, 0, \ldots, 0)$. It is now clear that E is a ring.

Now B is a finitely generated R-module by 6.5. Let $x_1,\ldots,x_n$ be the entries in a set of R-generators of B. It is clear that these entries generate E as an R-module.

Proof of 6.32. Let c be the conductor of B to A in Σ: $c = \{x \text{ in } \Sigma \mid xB \subseteq A\}$ is a left ideal in A. If $c = At$ with t in A then t is a unit in Σ because $c \cap R \neq 0$ (6.25). Since $cB \subseteq c$, we have $tB \subseteq At$, implying that $B \subseteq t^{-1}At$; hence $B = t^{-1}At$ by maximality of B. It thus suffices to prove that <u>when A is a central separable algebra over a DVR, every left ideal of A is principal</u>. We develop some facts needed to show this.

Let m be the maximal ideal of R. The Jacobson radical of A, $\mathcal{M}$, is a two-sided ideal, hence equal to IA for some ideal I of R (2.11), hence $\mathcal{M} \subseteq mA$. Conversely $mA \subseteq \mathcal{M}$ because A is a finitely generated R-module (1.10). Thus we have:

(1) $$\mathcal{M} = mA.$$

Let I be any left ideal of A. Both I and A are finitely generated torsion-free modules over the PID R, hence free. It follows from 1.13 that $\text{rank}_R I \leq \text{rank}_R A$, hence:

(2) $$\dim_{R/m} I/mI \leq \dim_{R/m} A/mA.$$

By Nakayama's lemma (1.9) the minimal number of generators of I as a left A-module equals the minimal number of generators of $I/\mathcal{M}I$ over $A/\mathcal{M}$, and by (1), equals the minimal number of generators of I/mI over A/mA. It suffices therefore to prove that I/mI is cyclic as an A/mA-module. The latter ring is a central simple R/m-algebra (2.3; Chapter 3), and in particular a simple Artinian ring; it therefore decomposes as $A/mA = J \oplus \ldots \oplus J$ (say s copies), J the unique indecomposable A/mA-module. I/mI is also a sum of (say t) copies of J. By (2), $t \leq s$. Thus I/mI is isomorphic to an ideal which is a direct summand of A/mA, hence principal. This completes the proof.

As an application of the preceding material, we compute the Brauer group of the ring of integers in an algebraic number field.

For the rest of this chapter we fix the following notation: K is an algebraic number field, i.e., a finite extension of $\mathbb{Q}$; R is its ring of integers, viz. the integral closure of $\mathbb{Z}$ in K; $\mathbb{P}$ is the set of "finite places" of K, i.e., non-zero (equivalently, height one) prime ideals of R; R_p is the localization of R at p, for p in $\mathbb{P}$ (hence R_p is a DVR contained in K); $\hat{R}_p$ is the completion of R_p, or equivalently the integral closure of R_p in the p-adic completion K_p of K.

The basic fact is:

6.34 <u>Proposition</u>. <u>For each</u> p <u>in</u> $\mathbb{P}$,
$0 \to B(R_p) \to B(K) \to B(K_p)$ <u>is exact</u>.

<u>Proof</u>. $B(R_p) \to B(K)$ is injective by 6.19. To see that the composition $B(R_p) \to B(K) \to B(K_p)$ is zero, observe that it coincides with the map $B(R_p) \to B(\hat{R}_p) \to B(K_p)$. We claim $B(\hat{R}_p) = 0$. We saw this for $K = \mathbb{Q}$ in 5.8 (g), and the same proof goes through in general: $\hat{R}_p$ is a complete local ring whose residue class field F is finite; $B(\hat{R}_p) \to B(F)$ is injective by 5.7 and $B(F) = 0$ by 3.7.

It remains to show that if Σ is a central simple K-algebra with $K_p \otimes_K \Sigma = \hat{\Sigma} \sim 1$ in $B(K_p)$, then $[\Sigma]$ in $B(K)$ lies in (the image of) $B(R_p)$. Let A be a maximal R_p-order in Σ. If we show A is central separable we are done, since $K \otimes_{R_p} A \simeq KA = \Sigma$ (6.16). A is central as before: any element in the center of A is both in the center K of Σ, and integral over R_p, which is integrally closed. To see that A is R_p-separable it suffices by 4.12 to see that $A/\mathfrak{m}A$ is $R_p/\mathfrak{m}$-separable, where $\mathfrak{m} = pR_p$ is the maximal ideal of R_p. The ($\mathfrak{m}$-adic) completion $\hat{R}_p$ of R_p is local with maximal ideal $\hat{\mathfrak{m}} = p\hat{R}_p$, and $\hat{R}_p/\hat{\mathfrak{m}} \simeq R_p/\mathfrak{m}$ (5.8 (c)). It follows that $\hat{A}/\hat{\mathfrak{m}}\hat{A} \simeq A/\mathfrak{m}A$, where $\hat{A} = \hat{R}_p \otimes_{R_p} A$. It therefore suffices, by 2.3, to see that $\hat{A}$ is $\hat{R}_p$-separable.

For this we need two further facts: $\hat{R}_p$ is a DVR, because R_p was (see BOURBAKI [2], Chapter 6, §5, Prop. 5), and $\hat{A}$ is a maximal $\hat{R}_p$-order in $\hat{\Sigma}$ (see DEURING [2], chapter VI ; proofs can also be found in AUSLANDER-GOLDMAN [1], Prop. 2.5, and REINER [1], §5). Now $\hat{\Sigma}$ is by hypothesis the endomorphism ring of a finite-dimensional K_p-vector space V. Let E be an $\hat{R}_p$-lattice in V; then $\text{End}_{\hat{R}_p}(E)$ is an $\hat{R}_p$-order in $\hat{\Sigma}$ (6.6). E is finitely generated and torsion-free as $\hat{R}_p$-module, hence free ($\hat{R}_p$ is a DVR, hence a PID). Thus $\text{End}_{\hat{R}_p}(E)$ is a central separable $\hat{R}_p$-algebra by 2.17. But 6.32 implies $\hat{A} \simeq \text{End}_{\hat{R}_p}(E)$, and we are done.

6.35 <u>Corollary</u>. The sequence

$$0 \to B(R) \to B(K) \to \bigoplus_{p \text{ in } \mathbb{P}} B(K_p),$$

where all maps are the obvious ones, is exact.

Proof. 6.34 gives an exact sequence for each p. Putting them all together yields an exact sequence

$$0 \to \bigcap_{p \text{ in } \mathbb{P}} B(R_p) \to B(K) \to \bigoplus_{p \text{ in } \mathbb{P}} B(K_p) \ .$$

(The right-hand map lands in the direct sum by 3.12 ff.) Now plug in 6.33 !

We can now prove:

6.36 <u>Theorem</u>. <u>Let</u> R <u>be the ring of integers in the algebraic number field</u> K, <u>and let</u> r <u>be the number of embeddings</u> $K \to \mathbb{R}$. Then:

(a) <u>If</u> $r = 0$ (K is "totally imaginary") <u>or</u> $r = 1$ (e.g., $K = \mathbb{Q}$, $R = \mathbb{Z}$) <u>then</u> $B(R) = 0$.

(b) <u>If</u> $r > 1$, $B(R)$ <u>is the direct sum of</u> $r - 1$ <u>copies of</u> $\mathbb{Z}/2\mathbb{Z}$.

Proof. We have, by 3.12 ff, an exact sequence

$$0 \to B(K) \to G \oplus \left(\bigoplus_{p \text{ in } \mathbb{P}} B(K_p) \right) \xrightarrow{g} \mathbb{Q}/\mathbb{Z} \to 0$$

where G is the direct sum of r copies of $\mathbb{Z}/2\mathbb{Z}$. Combining this with 6.35 yields an exact sequence

$$0 \to B(R) \to G \xrightarrow{g} \mathbb{Q}/\mathbb{Z}$$

in which $\sigma(x) = 0$ if and only if the number of non-zero components of x is <u>even</u>. In particular $B(R) = 0$ if $r = 0$. For $r \geq 1$, observe that exactly half of the elements of G have an even number of non-zero components (proof by induction on r). Thus

$B(R) \simeq \text{Ker}(\sigma)$ is a group of order 2^{r-1} and exponent 2, as claimed.

We remark that the above arguments and results for global fields of characteristic 0 (i.e., algebraic number fields) all remain valid, _mutatis mutandis_, for global fields of characteristic $p > 0$ (e.g., function fields over finite fields). See, for example, FOSSUM [1], Chapter 3.

We conclude this chapter with the following very remarkable corollary:

6.37 <u>Hauptsatz</u>. 2 <u>is prime</u>.

<u>Proof</u>. "Ecrasons, en effet, cette mouche avec un gros pavé." * Let $\mathbb{P}$ (resp. $\mathbb{P}'$) denote the set of all primes (resp. all <u>odd</u> primes). By 6.33 we have $\bigcap_{p \text{ in } \mathbb{P}} B(\mathbb{Z}_{(p)}) = B(\mathbb{Z})$. Hence 6.36 (a) implies that $\bigcap_{p \text{ in } \mathbb{P}} B(\mathbb{Z}_{(p)}) = 0$. On the other hand, we saw in 5.8 (e) that $\bigcap_{p \text{ in } \mathbb{P}'} B(\mathbb{Z}_{(p)}) \neq 0$. Consequently, $\mathbb{P}' \subseteq \mathbb{P}$ is a <u>proper</u> inclusion, or, otherwise stated, <u>there exists at least one even prime</u>. It is readily verified that no even integer greater than 2 is prime. The theorem follows.

---

* P. Samuel, Théorie Algébrique des Nombres, Hermann (Paris) 1967, p. 95 . In fact the proof will use <u>several</u> very "gros pavé" s , but no matter.

CHAPTER 7. GALOIS COHOMOLOGY.

In this chapter we define Galois extensions of rings and obtain the basic facts about the first and second cohomology of the Galois group. These results are used in chapter 8 to study $B(R) \to B(R[X])$.

Let G be a finite group of automorphisms of a commutative ring S , and let R be the subring $S^G = \{x \text{ in } S \mid \sigma x = x \text{ for all } \sigma \text{ in } G\}$. We say S **is a Galois extension of** R **with respect to** G if the following holds:

(7.1) **There are elements** $x_1,\ldots,x_m$, $y_1,\ldots,y_m$ **in** S **with** $\sum_{i=1}^{m} x_i y_i = 1$ **and** $\sum_{i=1}^{m} x_i(\sigma y_i) = 0$ **for all** $\sigma \neq 1$ **in** G .

If G is a finite group of ring automorphisms of a _field_ S and $R = S^G$, then S is a Galois extension of R , with Galois group G , in the usual sense (see LANG [1], VIII §1). It is easy to show that 7.1 is always satisfied in the field case; see the Remark following 7.3 below. Thus the definition above coincides with the usual notion, for fields.

A "fundamental theorem of Galois theory", providing a bijective correspondence between subgroups of G and subextensions of $S \supseteq R$ is available in our context. We will not develop this here; the interested reader can consult CHASE, HARRISON & ROSENBERG [1], §2 or DEMEYER & INGRAHAM [1], III §1 . Instead we turn to the main results of Galois cohomology in this setting, after recording some basic properties of Galois extensions. Since these basic results (7.3 and 7.4) are readily available, we will abbreviate the proofs by sending the reader to the standard sources (CHASE, HARRISON &

ROSENBERG [1] or DEMEYER & INGRAHAM [1]) at a few crucial points.

7.2 <u>Definitions</u>. Let S be a Galois extension of R with respect to G .

(a) A 1-<u>cocycle</u> is a map $h : G \to U(S)$ satisfying $h(\sigma\tau) = (h\sigma)\sigma(h\tau)$ for all σ, τ in G . A 1-<u>coboundary</u> is a map $h : G \to U(S)$ of the form $h\sigma = \sigma u/u$ for some u in $U(S)$. $H^1(G,U(S))$ is the quotient group {1-cocycles}/{1-coboundaries} .

(b) A 2-<u>cocycle</u> is a map $f : G \times G \to U(S)$ such that $f(\sigma\tau,\rho)f(\sigma,\tau) = \sigma f(\tau,\rho)f(\sigma,\tau\rho)$ for all σ, τ, ρ in G . A 2-<u>coboundary</u> is a map $f : G \times G \to U(S)$ of the form $f(\sigma,\tau) = h(\sigma\tau)/(h\sigma)\sigma(h\tau)$ for some map $h : G \to U(S)$; we write $f = \delta h$, and if f and g are cocycles with fg^{-1} a coboundary we write $f \sim g$. $H^2(G,U(S))$ is the quotient group {2-cocycles}/{2-coboundaries} .

(c) When f is a 2-cocycle, $\Delta(f,S,G)$ denotes the crossed product as defined in chapter 3 : as an S-module it is free on a basis $\{u_\sigma\}$ indexed by the elements of G , and the multiplication is given by $su_\sigma tu_\tau = s(\sigma t)f(\sigma,\tau)u_{\sigma\tau}$ (s,t in S) .

<u>Exercise</u>. The fact that f is a cocycle makes the multiplication in $\Delta(f,S,G)$ associative; $\Delta(f,S,G)$ is an R-algebra with $1 = f(1,1)^{-1}u_1$.

7.3 <u>Proposition</u>. Let S <u>be a Galois extension of</u> R <u>with respect to</u> G . Then:

(a) S <u>is faithfully projective as</u> R-<u>module, and its rank is constant and equal to the order of</u> G . (Definition of rank follows 1.12). R <u>is an</u> R-<u>direct summand of</u> S .

(b) S <u>is a separable</u> R-<u>algebra</u>.

(c) <u>The trace map</u> $\text{tr}: S \to R$ <u>defined by</u> $\text{tr}(s) = \sum_{\sigma \text{ in } G} \sigma s$ <u>is surjective</u>.

(d) <u>For any</u> R-<u>algebra</u> T, <u>let</u> G <u>act on</u> $T \otimes_R S$ <u>by</u> $\sigma(t \otimes s) = t \otimes \sigma s$. <u>Then</u> $T \otimes S$ <u>is a Galois extension of</u> $T \otimes R = T$ <u>with respect to</u> G.

(e) <u>Given</u> $\sigma \neq 1$ <u>in</u> G <u>and a maximal ideal</u> m <u>of</u> S, <u>there exists</u> s <u>in</u> S (<u>depending on</u> σ <u>and</u> m) <u>with</u> $\sigma s - s$ <u>not in</u> m.

<u>Remark</u>. (e) characterizes Galois extensions: if $R = S^G$ and (e) holds then S is a Galois extension of R with respect to G. See CHASE, HARRISON & ROSENBERG [1], (f) $\Rightarrow$ (b) of theorem 1.3.

<u>Proof</u>. Given x_i and y_i as in 7.1, define $f_i : S \to R$ by $f_i(s) = \sum_{\sigma \text{ in } G} \sigma(sy_i)$. Then $\{x_i, f_i\}$ is a projective basis (1.1) for S over R. Hence S is faithfully R-projective, and R is therefore a direct summand by 1.4. Now let E denote the direct product of copies of S, indexed by the elements of G. It can be shown (see either reference) that the map h from $S \otimes_R S$ to E given by sending $s \otimes t$ to $(\ldots, s(\sigma t), \ldots)$ is a well-defined isomorphism. The statement about the rank is an easy consequence: after localizing, $S \otimes S$ has rank $[S:R]^2$ whereas E has rank $n[S:R]$ where n is the order of G. This proves (a).

For (b), let $h : S \otimes S \to E$ be as before, and let p_1 be the projection of E onto S_1, the copy of S corresponding to the trivial element 1 of G. Use h (resp. $p_1 h$) to view E (resp. S_1) as $S \otimes S$-module. Since $p_1 h$ sends $s \otimes t$ to st, R-separability of S just means $S \otimes S$-projectivity of S_1. But the latter is clear, since S_1 is a direct summand of the free $S \otimes S$-module E.

(c) Since tr is R-linear, tr(S) = I is an ideal of R. Since $\sum_i x_i \text{tr}(y_i) = 1$ for x_i, y_i as in 7.1, we have IS = S. 1.2 therefore supplies a in I with (1-a)S = 0. But a can only be 1, hence 1 is in I and I = R.

(d) If $\{x_i, y_i\}$ satisfy 7.1 for S over R then $\{1 \otimes x_i, 1 \otimes y_i\}$ satisfy 7.1 for T ⊗ S over T. Hence we need only show that $T \subseteq (T \otimes S)^G$ is an equality. Let x be in $(T \otimes S)^G$ and use (c) to choose y in S with tr(y) = 1. Since 1 ⊗ tr maps T ⊗ S to T, x = x((1⊗tr)(1⊗y)) = (1⊗tr)(x(1⊗y)) is in T, q.e.d.

(e) If m is a maximal ideal with σs - s in m for all s in S, take x_i and y_i as in 7.1 to get the contradiction that $1 = \sum_i x_i(y_i - \sigma y_i)$ is in m.

7.4 Proposition. Let S be a Galois extension of R with respect to G. Then:

(a) The map j : Δ(1,S,G) → $\text{End}_R(S)$ defined by $j(au_\sigma)x = a(\sigma x)$ is an isomorphism of R-algebras.

(b) If f ~ g then Δ(f,S,G) and Δ(g,S,G) are isomorphic R-algebras.

(c) For any 2-cocycle f, S is a maximal commutative subalgebra of Δ(f,S,G).

Proof. (a) Clearly j is a homomorphism and lands in $\text{End}_R(S)$. Injectivity and surjectivity are easy computations which the reader can recreate or find in either reference.

(b) We have h : G → U(S) with g = f(δh). Define a map from Δ(f,S,G) to Δ(g,S,G) by $su_\sigma \mapsto s(h\sigma)u_\sigma$. This is certainly an isomorphism if it is a homomorphism, and the latter is an easy computation.

Before proving (c) we mention an important consequence of (b). Given a 2-cocycle f, define $h : G \to U(S)$ by $h\sigma = \sigma f(1,1)$. Then $f \sim f'= f(\delta h)$, and f' has the property that $f'(\sigma,1) = 1 = f'(1,\sigma)$ for all σ in G. Thus in forming crossed products we can (and will) always assume the cocycles are normalized so that $1 = u_1$.

(c) If su_σ commutes with tu_1 for all t in S we get $s(\sigma t - t) = 0$ for all t in S. If $\sigma \neq 1$ this contradicts 7.3(e) unless $s = 0$.

The following result is a partial converse to 7.4(b). It can be strengthened to give a genuine converse (i.e., the hypothesis $\phi|S = 1$ can be dropped) but we shall not prove this here.

7.5 <u>Lemma</u>. <u>If</u> $\phi = \Delta(f,S,G) \to \Delta(g,S,G)$ <u>is an</u> R-<u>algebra isomorphism which is the identity on</u> S, <u>then</u> $f \sim g$.

<u>Proof</u>. To avoid confusion, write $\Delta(f,S,G) = \oplus Su_\sigma$, $\Delta(g,S,G) = \oplus Sv_\sigma$. We are assuming, then, that $\phi(su_1) = sv_1$ for s in S. For s in S we have $\sigma s = u_\sigma s u_\sigma^{-1}$ in $\Delta(f,S,G)$, hence $\phi(u_\sigma) s \phi(u_\sigma)^{-1} = \phi(\sigma s) = \sigma s = v_\sigma s v_\sigma^{-1}$ in $\Delta(g,S,G)$. Thus $v_\sigma^{-1} \phi(u_\sigma)$ commutes with $Sv_1 = S$, hence is in S by 7.4(c). Thus $\phi(u_\sigma) = s_\sigma v_\sigma$ for some s_σ in $U(S)$. Define $h : G \to U(S)$ by $h(\sigma) = s_\sigma$. An easy computation shows that $g = f(\delta h)$:
$f(\sigma,\tau)s_{\sigma\tau}v_{\sigma\tau} = \phi(u_\sigma u_\tau) = (\phi u_\sigma)(\phi u_\tau) = s_\sigma \sigma(s_\tau) g(\sigma,\tau) v_{\sigma\tau}$.

Let S be a Galois extension of R with respect to G. If S is a field, the two basic results of Galois cohomology are that $H^1(G,U(S)) = 0$ (this is variously referred to as Noether's theorem, or Hilbert's theorem 90), and $H^2(G,U(S)) \simeq B(S/R)$, as sketched in chapter 3 . The simplest way to describe the generalizations of these results to Galois extensions of rings is as follows: <u>the result about H^1 survives if Pic(R) = 0 , and the result about H^2 survives if</u> Pic(S) = 0 .

(Pic(X) = 0 means every projective X-module of constant rank one is free. For the reader who knows more about Pic than this, we remark that the two cohomology groups in question fit into an exact sequence whose other terms involve Pic(R) , Pic(S) , and B(S/R) , from which the two results indicated can be extracted in the appropriate special cases. See DEMEYER & INGRAHAM [1], Chapter IV.)

The theorem about H^1 is by far the easier of the two. We prove in 7.6 that $H^1(G,U(S))$ vanishes under a hypothesis which looks much weaker than Pic(R) = 0 .

The traditional link between $H^2(G,U(S))$ and B(S/R) (in DEMEYER & INGRAHAM [1], AUSLANDER & GOLDMAN [2], CHASE & ROSENBERG [1], ROSENBERG & ZELINSKY [1],...) is the one indicated in chapter 3 for the field case: given a cocycle, build a crossed product. We have, instead, resurrected an idea which goes back to DIEUDONNÉ [2], to construct an isomorphism going the other way: from B(S/R) to $H^2(G,U(S))$. The proof (in 7.12) that our map is an isomorphism when Pic(S) = 0 is more conceptual than the corresponding proof for the traditional map. (The crucial thing in either approach is to show that the map in question is a homomorphism, and the proof of this for the map $H^2(G,U(S)) \to B(S/R)$ is a truly horrible cocycle computation.) The proof of 7.12 shows that our map is the inverse of the traditional map $H^2(G,U(S)) \to B(S/R)$.

Recall that an automorphism ϕ of an R-algebra A is <u>inner</u> if $\phi x = uxu^{-1}$ for all x in A , for some unit u of A . <u>If</u>

Pic(R) vanishes, every endomorphism of every central separable R-algebra is an inner automorphism. See, for example, BASS [1], III, 4.4 and 4.6 , or DEMEYER & INGRAHAM [1], II, 6.1 and 6.2 ; this result is a general form of the Skolem-Noether theorem, 3.6 . In particular, then, the following applies whenever Pic(R) = 0 :

7.6 Theorem. Let S be a Galois extension of R with respect to G , and assume that every R-algebra automorphism of $End_R(S)$ (or equivalently of $\Delta = \Delta(1,S,G)$) is inner. Then $H^1(G,U(S)) = 0$.

Proof. Let $h : G \to U(S)$ be a 1-cocycle, and define $\phi = \phi_h : \Delta \to \Delta$ by $\phi(su_\sigma) = (h\sigma)su_\sigma$. It follows from the cocycle condition that $h(1) = 1$ so that ϕ is the identify on $S = Su_1$, and that ϕ is an R-algebra automorphism. By hypothesis there is a unit v in Δ with $(h\sigma)su_\sigma = v(su_\sigma)v^{-1}$ for all s in S , σ in G . Since S is a maximal commutative subalgebra of Δ (7.4(c)) the fact that ϕ , an inner automorphism by v , is the identity on S shows that v is in $U(S)$. Thus $(h\sigma)su_\sigma = v(su_\sigma)v^{-1} = v(\sigma v^{-1})su_\sigma$, and $h\sigma = \sigma u/u$ for $u = v^{-1}$; done.

7.7 Proposition. Let A be a central separable R-algebra, S a maximal commutative subalgebra of A . Viewing A as a right S-module, the map $f : A \otimes_R S \to End_S(A)$ given by $f(a\otimes s)(x) = axs$ is an S-algebra isomorphism. If A is S-projective then S is R-separable and splits A . If S is R-separable then A is S-projective, and projective as a left $A \otimes_R S$-module.

Proof. Let $\eta : A \otimes_R A^\circ \to End_R(A)$ be the canonical isomorphism defined by $\eta(a\otimes b^\circ)(x) = axb$. Then $End_S(A) \cong (A\otimes A^\circ)^{1\otimes S} = A \otimes S$ (Use 2.3(a) and the fact that $(A^\circ)^S = S$). The statement about f

follows readily. If A is S-projective, then 4.14(b) implies that S is R-separable. If S is R-separable then A is S-projective by 2.4 . Then $A \otimes A^o$ is $A \otimes$ S-projective, and since A is $A \otimes A^o$-projective, the last statement follows.

7.8 <u>Lemma</u>. <u>Let</u> S <u>be a Galois extension of</u> R <u>with respect to</u> G , <u>and let</u> $f : G \times G \to U(S)$ <u>be a</u> 2-<u>cocycle</u>. <u>Then</u> $\Delta(f,S,G)$ <u>is a central separable</u> R-<u>algebra split by</u> S , <u>and containing</u> S <u>as a maximal commutative subalgebra</u>.

<u>Proof</u>. The last statement was checked in 7.4 (c) . Write Δ for $\Delta(f,S,G)$. Define a map $j : \Delta \otimes_R S \to End_S(\Delta)$ (S acts on Δ on the right) by $j(su_\sigma \otimes t)(xu_\tau) = s\sigma(x)f(\sigma,\tau)\sigma\tau(t)u_{\sigma\tau}$. One can check that j is a well-defined S-algebra homomorphism. To show j is onto we note that $End_S(\Delta)$ is a free S-module generated by all $e_{\alpha,\beta}$ (α,β in G) where $e_{\alpha,\beta}(u_\sigma) = \delta_{\alpha,\sigma}u_\beta$. Let $x_1,\ldots,x_m$, $y_1,\ldots,y_m$ in S satisfy $\Sigma x_i \sigma y_i = \delta_{1,\sigma}$ (7.1) ; then $e_{\alpha,\beta} = j(z_{\alpha,\beta})$ where

$$z_{\alpha,\beta} = \sum_{i,\gamma} \delta_{\gamma,\alpha^{-1}\beta} f(\gamma,\alpha)^{-1} x_i u_\gamma \otimes \beta^{-1}(y_i) .$$

Since $\Delta \otimes_R S$ and $End_S(\Delta)$ are free S-modules of rank n^2 (n = $|G|$) , j being onto implies j is an isomorphism (1.14) . Thus $\Delta \otimes_R S$ is a central separable S-algebra. It follows from 2.14 that Δ is a central separable R-algebra (Exercise) .

7.9 <u>Lemma</u>. <u>Let</u> A <u>be a central separable</u> R-<u>algebra</u>, S <u>a maximal commutative subalgebra of</u> A <u>which is a Galois extension of</u> R <u>with group</u> G . <u>Assume</u> Pic(S) = 0 . <u>Then each</u> σ <u>in</u> G <u>extends to an inner automorphism of</u> A .

Proof. Let $\bar{A} = \{\bar{x} \mid x \text{ in } A\}$ with left $A \otimes S$-action $(a \otimes s)(\bar{x}) = \overline{ax\sigma(s)}$. By 7.7 the map $j : A \otimes S \to \text{End}_S(A)$, given by $j(a \otimes s)(x) = axs$, is an S-algebra isomorphism. Give $\bar{A}$ a left $\text{End}_S(A)$-module structure by $\alpha \bar{x} = \overline{j^{-1}(\alpha)x}$. Give $\bar{A}$ a right S-module structure by $\bar{x}s = \overline{x\sigma(s)}$. View S as embedded in $\text{End}_S(A)$ by $s \to R_s$, where $R_s(x) = xs$. Then $\bar{A}$ is an $(\text{End}_S(A), S)$-bimodule, and the left S-action arising from the embedding of S in $\text{End}_S(A)$ agrees with the right S-action defined above.

Since S is R-separable (7.3(b)), A is a faithfully projective right S-module (7.7), hence there exists a left S-module N and an $\text{End}_S(A)$-isomorphism $f: \bar{A} \to A \otimes_S N$. By the remark just preceding, f is an S-module isomorphism as well.

We wish to show N is a faithfully projective S-module of rank 1. $\bar{A}$ is $A \otimes S$-projective since it is R-projective and $A \otimes S$ is R-separable (2.3(d), 2.4); hence by 1.18, N is S-projective. But N is then faithfully projective because $\bar{A}$ is S-faithful. To conclude that N has rank 1 over S it suffices to show that A and $\bar{A}$ have the same rank at each prime P of S, and to do this it suffices to show that A has constant rank, since the rank of $\bar{A}$ at P equals the rank of A at σP (Exercise). Now S has constant rank $n = |G|$ as an R-module (7.3(a)) and $A \otimes S \simeq \text{End}_S(A)$. For P a prime of S let $p = P \cap R$. It is an easy exercise to verify the following equalities: $\text{rk}_p(A)^2 = \text{rk}_p(\text{End}_S(A)) = \text{rk}_p(A \otimes S) = \text{rk}_p(A \otimes_S (S \otimes S)) = \text{rk}_p(A) \cdot \text{rk}_p(S \otimes S) = \text{rk}_p(A) \cdot \text{rk}_p S = \text{rk}_p(A)n$. Hence $\text{rk}_p(A) = n$ for all primes P of S, and N is projective of rank 1.

If $\text{Pic}(S) = 0$ then $N \simeq S$ and there is an $\text{End}_S(A)$-module isomorphism, hence an $A \otimes S$-module isomorphism, $h : \bar{A} \to A$. We have $h(\overline{ax\sigma(s)}) = ah(\bar{x})s$ for s in S, x, a in A. Let $h(\bar{v}) = 1$, $h(\bar{1}) = u$. Choosing $a = v$, $x = s = 1$, we obtain $1 = vu$; then $h(\overline{uv}) = h(\bar{1})$ follows, hence $uv = 1$ and u is a unit. Putting $a = x = 1$ yields that $\sigma(s) = usu^{-1}$.

7.10 <u>Proposition</u>. <u>Let</u> S <u>be a Galois extension of</u> R <u>with respect to</u> G <u>and let</u> A <u>be a central separable</u> R-<u>algebra containing</u> S <u>as a maximal commutative subalgebra</u>. <u>Then</u> $A \simeq \Delta(f,S,G)$ <u>for some cocycle</u> f <u>if and only if every element of</u> G <u>extends to an inner automorphism of</u> A .

<u>Proof</u>. Then "only if" part is clear: in $\Delta(f,S,G)$, $u_\sigma s = \sigma(s)u_\sigma$ for s in S , hence σ extends to inner automorphism by u_σ . Also, the units u_σ determine f , since $u_\sigma u_\tau = f(\sigma,\tau)u_{\sigma\tau}$. These observations show how to prove the converse: For each σ in G we have a unit t_σ in A with $\sigma s = t_\sigma s t_\sigma^{-1}$ for s in S ; choose $t_1 = 1$. Then let $f(\sigma,\tau)$ be the unit $t_\sigma t_\tau t_{\sigma\tau}^{-1}$ of A . For any s in S we have $f(\sigma,\tau)sf(\sigma,\tau)^{-1} = t_\sigma t_\tau t_{\sigma\tau}^{-1} s t_{\sigma\tau} t_\tau^{-1} t_\sigma^{-1} = \sigma(\tau((\sigma\tau)^{-1}s)) = s$, hence, since S is a maximal commutative subalgebra, $f(\sigma,\tau)$ is a unit of S . It is easily checked that f is a cocycle: $f(\sigma,\tau)f(\sigma\tau,\rho) = t_\sigma t_\tau t_\rho t_{\sigma\tau\rho}^{-1} = \sigma f(\tau,\rho)f(\sigma,\tau\rho)$. Now let Δ be the crossed product $\Delta(f,S,G)$ and define $h : \Delta \to A$ by $h(u_\sigma) = t_\sigma$. Since we chose $t_1 = 1$, h is the identity on S , and is clearly an R-algebra homomorphism. It is injective since A is central separable (2.12(b)) , and $A = h(\Delta) \otimes A^{h(\Delta)}$ by 2.13 . But clearly $A^{h(\Delta)} \subseteq A^{h(S)} = S$, so that $A^{h(\Delta)} \subseteq S^{h(\Delta)} \subseteq h(\Delta^\Delta) = R$. Hence $h(\Delta) = A$, q.e.d.

7.11 <u>Proposition</u>. <u>Let</u> S <u>be an</u> R-<u>algebra which is a faithfully projective</u> R-<u>module</u>. <u>Let</u> A <u>be a central separable</u> R-<u>algebra split by</u> S , <u>and let</u> $h : A \otimes_R S \to \mathrm{End}_S(E)$ <u>be an</u> S-<u>algebra isomorphism</u>, <u>with</u> E <u>a faithfully projective right</u> S-<u>module</u>. <u>View</u> E <u>as a left</u> A-<u>module via</u> $ax = h(a \otimes 1)(x)$. <u>Then</u>:

(a) $\text{End}_A(E)$ is a central separable R-algebra which is equivalent to A°, i.e., $[\text{End}_A(E)] = [A^\circ]$ in $B(R)$.

(b) Viewing $\text{End}_A(E)$ as contained in $\text{End}_R(E)$, $h(1 \otimes S)$ is a maximal commutative subalgebra of $\text{End}_A(E)$.

(c) Assume $\text{Pic}(S) = 0$. Let $[A] = [B]$ in $B(R)$ and let $j : B \otimes_R S \to \text{End}_S(F)$ be an S-algebra isomorphism, with F a faithfully projective S-module. Then there exists an R-algebra isomorphism $f : \text{End}_A(E) \to \text{End}_B(F)$ with $fh(1 \otimes s) = j(1 \otimes s)$.

Proof. (a) A is embedded in $\text{End}_R(E)$ via its action on E. Then $\text{End}_A(E) = \text{End}_R(E)^A$. By 2.13, $A \otimes \text{End}_A(E) \simeq \text{End}_R(E)$, hence $\text{End}_A(E)$ is a central separable R-algebra equivalent to A° (2.15).

(b) For s in S, $h(1 \otimes s)$ is right multiplication by s on E since h is an S-map. E is an (A,S)-bimodule and $h(1 \otimes s)$ is in $\text{End}_A(E)$. We have $\text{End}_A(E)^{h(1 \otimes S)} \subseteq \text{End}_S(E)^{h(A \otimes 1)} = h((A \otimes S)^A)$. By 2.3(a), $(A \otimes S)^A = 1 \otimes S$, hence $h(1 \otimes S)$ is a maximal commutative subalgebra of $\text{End}_A(E)$.

(c) First let $A = B$ and let h, j, E, F be as described. Put a left $\text{End}_S(E)$-module structure on F by $\alpha x = jh^{-1}(\alpha)(x)$. By the Morita theory (1.18), the correspondence $X \mapsto E \otimes_S X$ yields a category isomorphism S-Mod $\leftrightarrow \text{End}_S(E)$ - Mod. Hence $F \simeq E \otimes_S N$ as $\text{End}_S(E)$-modules for some S-module N. Now $h(1 \otimes S) \subseteq \text{End}_S(E)$, and $h(1 \otimes s)(x) = xs$ for x in E; similarly, for y in F, $h(1 \otimes s)(y) = jh^{-1}h(1 \otimes s)(y) = j(1 \otimes s)(y) = ys$. Thus $F \simeq E \otimes_S N$ as S-modules as well. Since F is faithfully projective as an $\text{End}_S(F)$-module (1.18), hence as an $A \otimes S$-module, and the $\text{End}_S(E)$ structure on F comes via jh^{-1}, it follows that F is a faithfully projective $\text{End}_S(E)$-module. Thus N is a faithfully projective S-module (1.18). Being faithfully projective, E and F have nonzero rank at each prime p of S. They have the same rank at each p since $S_p \otimes_S \text{End}_S(X) \simeq \text{End}_{S_p}(S_p \otimes_S X)$ for $X = E, F$ (1.6). It follows that N has rank 1. If $\text{Pic}(S) = 0$ then $N \simeq S$ as S-modules and $E \simeq F$ as $(\text{End}_S(E), S)$-bimodules. Now A embeds in $\text{End}_S(E)$ as $h(A \otimes 1)$, and the A-structure on F induced by the restriction of the $\text{End}_S(E)$-structure agrees with the one defined by $ax = j(a \otimes 1)(x)$. Thus $\text{End}_A(E) \simeq \text{End}_A(F)$ and it is clear that this isomorphism carries $h(1 \otimes S)$ to $jh^{-1}h(1 \otimes S) = j(1 \otimes S)$.

This shows that $\text{End}_A(E)$ is independent of E and h in the manner claimed in (c). To show that $\text{End}_A(E)$ depends only on $[A]$, it suffices to consider $B = \text{End}_R(P) \otimes A$, P a faithfully projective R-module. Take $1 \otimes h : \text{End}_R(P) \otimes A \otimes S \to \text{End}_R(P) \otimes \text{End}_S(E)$. Since $\text{End}_R(P) \otimes \text{End}_S(E)$ is naturally isomorphic to $\text{End}_S(P \otimes E)$ (1.5), we have that $\text{End}_B(P \otimes E) = \text{End}_R(P \otimes E)^{\text{End}(P) \otimes A} \simeq (\text{End}_R(P) \otimes \text{End}_R(E))^{\text{End}(P) \otimes A} = 1 \otimes \text{End}_R(E)^A$ (by 2.3(a)), and the last term equals $\text{End}_A(E)$. It is easily checked that the overall isomorphism $\text{End}_B(P \otimes E) \simeq \text{End}_A(E)$ carries $(1 \otimes h)(1 \otimes 1 \otimes s)$ to $h(1 \otimes s)$.

7.12 <u>Theorem</u>. <u>Let</u> S <u>be a Galois extension of</u> R <u>with respect to</u> G, <u>and assume</u> $\text{Pic}(S) = 0$. <u>Then there exists an isomorphism</u> $B(S/R) \simeq H^2(G, U(S))$.

<u>Proof</u>. Let $[A]$ be in $B(S/R)$ and let $h : A^o \otimes S \to \text{End}_S(E)$ be an S-algebra isomorphism, with E a faithfully projective right S-module. By 7.11, S (identified with $h(1 \otimes S)$) is a maximal commutative subalgebra of $B = \text{End}_{A^o}(E)$, and $[B] = [A]$ in $B(R)$. By 7.9 and 7.10, $B \simeq \Delta(f, S, g)$ for some cocycle $f : G \times G \to U(S)$, and by 7.5 and 7.11 the cohomology class of f in $H^2(G, U(S))$, $[f]$, is well-defined as a function of $[A]$. Set $\phi([A]) = [f]$.

Suppose $\phi([A_1]) = \phi([A_2])$. Let $h_i : A_i^o \otimes S \to \text{End}_S(E_i)$ be isomorphisms and let $B_i = \text{End}_{A_i^\circ}(E_i)$. Then $B_i \simeq \Delta(f_i, S, G)$ and $f_1 \sim f_2$. By 7.4(b), $B_1 \simeq B_2$ as R-algebras. But $[B_i] = [A_i]$, hence ϕ is one-one.

ϕ is onto as well. Let $f: G \times G \to U(S)$ be a cocycle. By 7.8, $\Delta(f,S,G) = A$ is a central separable R-algebra split by S and containing S as a maximal commutative subalgebra; in fact $A^\circ \otimes S \simeq \text{End}_S(A^\circ)$ by 7.7. Then $[g] = \phi([A])$ is such that $\Delta(g,S,G) \simeq \text{End}_{A^\circ}(A^\circ)$, which is just A, hence $[g] = [f]$ by 7.5, and ϕ is onto.

We now show that ϕ is a homomorphism. If we have isomorphisms $A^\circ \otimes S \simeq \text{End}_S(E)$, $B^\circ \otimes S \simeq \text{End}_S(F)$, then tensoring over S and using 1.5 we obtain an isomorphism $(A \otimes B)^\circ \otimes S \simeq \text{End}_S(E \otimes_S F)$. For an element s of S, write R_s for the map in $\text{End}_{A^\circ}(E)$ sending x to xs. By the foregoing work (7.9, 7.11) there are units u_σ in $\text{End}_{A^\circ}(E)$ such that $u_\sigma R_s u_\sigma^{-1} = R_{\sigma(s)}$ for each σ in G, s in S. Thus for each element x of E we have $u_\sigma(u_\sigma^{-1}(x)s) = x\sigma(s)$, or, writing $y = u_\sigma^{-1}(x)$:

(*) $\quad u_\sigma(ys) = u_\sigma(y)\sigma(s)$ for y in E, s in S, σ in G.

There are units v_σ in $\text{End}_{B^\circ}(F)$ satisfying a similar condition with respect to F. Using (*) it is straightforward to show that there are well-defined elements in $\text{End}_{(A \otimes B)^\circ}(E \otimes_S F)$, $w_\sigma = u_\sigma \otimes v_\sigma$, satisfying $w_\sigma(x \otimes y) = u_\sigma(x) \otimes v_\sigma(y)$ for x in E, y in F. Clearly w_σ is a unit in $\text{End}_{(A \otimes B)^\circ}(E \otimes_S F)$, and it is straightforward to compute that inner automorphism by w_σ extends the map $\sigma: S \to S$, i.e., that $w_\sigma R_s w_\sigma^{-1} = R_{\sigma(s)}$ where $R_s(x \otimes y) = xs \otimes y = x \otimes ys$ for x in E, y in F. It is clear that if $u_\sigma u_\tau = f(\sigma,\tau) u_{\sigma\tau}$ and $v_\sigma v_\tau = g(\sigma,\tau) v_{\sigma\tau}$, then $w_\sigma w_\tau = f(\sigma,\tau) g(\sigma,\tau) w_{\sigma\tau}$. Hence if $\phi([A]) = [f]$ and $\phi([B]) = [g]$ then

$\phi([A \otimes B]) = [fg]$, hence ϕ is a homomorphism. This completes the proof of 7.12 .

7.13 <u>Exercise</u>. (a) Let S be a commutative faithful R-algebra, finitely generated as R-module, where R is local. Let m be the maximal ideal of R and let $M_1,\ldots,M_t$ be the maximal ideals of S (cf. 4.3(b)) . Let $N = \bigcap M_i$. We have seen that $mS \subseteq N$ (4.3(a)) . Assuming S is R-separable, show that $mS = N$. (Outline: S/mS is a direct product $F_1 \times \ldots \times F_s$ of fields by 2.7 , and S/N is a direct product $G_1 \times \ldots \times G_t$ of fields, viz. $G_i = S/M_i$. The inclusion $mS \subseteq N$ yields a surjection $S/mS \to S/N$; since s = (number of maximal ideals of S/mS) = (number of maximal ideals of S containing mS) = t , it is an isomorphism.)

(b) Assume R is local and S is a Galois extension of R with respect to a group of order n . Use (a) to show that S has at most n maximal ideals.

(c) Globalize (b) : If $S \supseteq R$ is a Galois extension with respect to a group of order n , then for each prime ideal p of R there are at most n prime ideals P of S such that $P \cap R = p$.

CHAPTER 8. TSEN'S THEOREM AND $B(R) \to B(R[X])$.

The main goal of this chapter is Tsen's theorem, which states that $B(K) = 0$ if K is an algebraic extension of $k(X)$, where k is an algebraically closed field. The proof follows the exposition given in GREENBERG [1]; the reader is referred to this book for a broader view of the phenomena encountered. Tsen's theorem, and the Galois cohomology of chapter 7, are then applied to study $B(R) \to B(R[X])$.

The following properties of a field K are relevant:

C_0 : K is algebraically closed.

C_1 : Every form (homogeneous polynomial) in $K[X_1,\ldots,X_n]$ whose degree is less than n has a nontrivial zero.

8.1 **Proposition.** <u>If</u> K <u>is</u> C_0 <u>and</u> $f_1,\ldots,f_r$ <u>are forms in</u> $K[X_1,\ldots,X_n]$, <u>and</u> $r < n$ <u>then</u> $f_1,\ldots,f_r$ <u>have a common nontrivial zero in</u> K^n .

We shall use this result without proving it. It is essentially a result from algebraic geometry, as it states that the intersection of r hypersurfaces in $P^{n-1}(K)$ (projective $(n-1)$-space) is nonempty (see e.g., ZARISKI & SAMUEL [2], p. 209).

8.2 <u>Lemma.</u> <u>If</u> K <u>has property</u> C_1 , <u>then</u> $B(K) = 0$.

<u>Proof.</u> Let D be a central division algebra over K , of dimension n^2 . Choose a basis $a_1,\ldots,a_{n^2}$ of D over K . According to 6.1 , $N_{red} : D \to K$ determines a form $f(x_1,\ldots,x_{n^2})$ (where $x_1,\ldots,x_m$ can be thought of as the coordinates of an element $\Sigma x_i a_i$ of A). If $n > 1$, i.e., if D does not represent the trivial element of $B(K)$, f must have a nontrivial

zero in K^n. But N_{red} is multiplicative and D is a division algebra, so the last conclusion is impossible.

The aim will now be to establish that if k is algebraically closed, then any algebraic extension K of $k(X)$ satisfies C_1. The case $K = k(X)$ is disposed of first:

8.3 **Lemma.** *Let* k *be algebraically closed, and let* $K = k(X)$. *Then* K *satisfies* C_1, *hence* $B(K) = 0$.

Proof. Let f be a form in $K[X_1,\ldots,X_n]$ of degree $d < n$. We may assume f has coefficients in $k[X]$. Finding a nontrivial zero of f in $k[X]^n$ amounts to finding elements y_{ij} in k yielding polynomials

$$x_i = \sum_{j=0}^{m} y_{ij} X^j$$

which determine a zero $(x_1,\ldots,x_n)$ of f. Treat the y_{ij} as unknowns and write $k[Y]$ for $k[y_{1,0},\ldots,y_{n,m}]$. Substituting the above expression for x_i in f, and writing each coefficient of f as a polynomial in $k[X]$ (say of degree at most r) yields:

$$f(X) = f_0(Y) + f_1(Y)X + \ldots + f_{dm+r}(Y)X^{dm+r}$$

where $f_i(Y)$ is a form in the $n(m+1)$ variables $y_{1,0},\ldots,y_{n,m}$. For $m > (r-n+1)/(n-d)$ we have more variables than forms. By 8.1, there is a common nontrivial zero for the f_i. This determines a nontrivial zero $(x_1,\ldots,x_n)$ for f.

The following is Tsen's theorem. The special case 8.3 will suffice for our applications.

8.4 Theorem. *Let* k *be an algebraically closed field. Let* K *be an algebraic extension of* $k(X)$. *Then* $B(K) = 0$.

Proof. Because of 8.2 and 8.3, it suffices to show that if F is a C_1 field, so is any algebraic extension K of F. Because a form over K involves only a finite number of coefficients, we may assume that K is a *finite* extension of F. To simplify matters a bit, though, we shall use the following fact about the particular F before us, viz. $k(X)$:

8.5 Exercise. $F = k(X)$ admits algebraic extensions of any degree t. If E is an algebraic extension of degree t over F, then E determines a form ϕ in $F[X_1,\ldots,X_t]$ which has degree t and has no nontrivial zero in F^n. (Hint: Consider the determinant of multiplication by the elements of E.)

Now let f be a form in $K[X_1,\ldots,X_n]$ of degree $d < n$. We wish to show that f has a nontrivial zero. Let $w_1,\ldots,w_e$ be a basis of K over F. Following the same idea as in the proof of 8.3, we seek elements y_{ij} in F, not all zero, for which

$$x_i = \sum_{j=1}^{e} y_{ij} w_j \qquad i = 1,\ldots,n$$

yields a solution (necessarily nontrivial) of f. Substitute these expressions for the x_i into f, and write the coefficients of f as linear combinations of $w_1,\ldots,w_e$ with coefficients in F. Various products of powers of the w_i arise, but these can again be expressed as linear combinations of $w_1,\ldots,w_e$. Thus we obtain an expression

$$f = f_1(y)w_1 + \ldots + f_e(y)w_e$$

where $f_i(y)$ is a form in $F[y_{11},\ldots,y_{ne}]$ of the same degree d as f.

The essential ingredients now are that $f_1,\ldots,f_e$ are e forms over F, each in ne variables, of degree at most d, and $d < n$. We shall show that this data implies the existence of a nontrivial zero for f. Say t is an integer $> e$, whose precise value will be decided later. Let ϕ in $k[X_1,\ldots,X_t]$ be a form of degree t having no nontrival zero (8.5). Substitute $f_1,\ldots,f_e$ for the X_i in the following fashion:

$$\psi = \phi(f_1,\ldots,f_e|f_1,\ldots,f_e|\ldots|f_1,\ldots,f_e|0,\ldots,0) :$$

as many f_i as possible are inserted, and the variables used in each group of f_i's are taken distinct from those in the other groups. The number of variables now introduced is $ne[t/e]$ ([ ] = greatest integer function), and the degree of ψ in these variables is at most dt. If ψ were to have a nontrivial zero, then $f_1,\ldots,f_e$ would vanish simultaneously at this point, since ϕ has no nontrivial zeroes. Now ψ has coefficients in F, which is C_1. Hence for ψ to have a nontrivial zero it suffices to arrange that $dt < ne[t/e]$, or better yet, that $de([t/e]+1) < ne[t/e]$. Since $d < n$ this can be done by choosing t large. The proof of 8.4 is therefore complete.

We turn now to the behavior of the Brauer group under polynomial extension, i.e., to the study of the map $B(R) \to B(R[X])$ induced by the inclusion $i : R \to R[X]$. i is split by the homomorphism $j : R[X] \to R$, $j(X) = 0$. Therefore $B(i)$ is injective, and $B(R[X]) \cong B(R) \oplus B'(R[X])$, where $B'(R[X]) = \ker B(j)$. Hence $B(i)$ is an isomorphism $\iff B(i)$ is surjective $\iff B(j)$ is injective $\iff B(j)$ is an isomorphism $\iff B(i)B(j)$ = Identity on

$B(R[X])$. The question of which rings R satisfy these conditions, i.e., for which R is $B(R) \to B(R[X])$ an isomorphism, seems to be virtually untouched except when R is a field.

8.6 **Proposition.** *Let A be a central separable $K[X]$-algebra where K is a field. Then the following are equivalent:*

(1) *$[A]$ is in the image of $B(K) \to B(K[X])$.*

(2) *There is a finite separable field extension L of K such that $L[X]$ splits A.*

(3) *There is a finite Galois field extension L of K such that $L[X]$ splits A.*

Proof. The equivalence of (2) and (3) is clear. If (1) holds, there is a central separable K-algebra B with $[K[X] \otimes_K B] = [A]$ in $B(K[X])$. B is split by some finite separable field extension L of K (see chapter 3), and then clearly $L[X]$ splits A. Conversely, assuming (3), we have that $[A]$ is in $B(S/R)$ where $S = L[X]$ is a Galois extension of $R = K[X]$, by 7.3 (d). Since S is a PID we have $\mathrm{Pic}(S) = 0$ (indeed all projectives are free), hence by 7.12 there is a cocycle $f : G \times G \to U(S)$ such that $[A] = [\Delta(f,S,G)]$ in $B(R)$. But $U(S) = U(L[X]) = U(L)$, so that we can view f as a cocycle $G \times G \to U(L)$ and form the central separable K-algebra $\Delta(f,L,G)$. Clearly $K[X] \otimes_K \Delta(f,L,G) \cong \Delta(f,L[X],G)$, which shows that $[A] = [K[X] \otimes_K \Delta(f,L,G)]$ is in the image of $B(K) \to B(K[X])$.

Recall that a field K is called **perfect** if the homomorphism $K \to K$, $x \to x^p$, is surjective, where p is the characteristic exponent of K, viz. $\max(1, \mathrm{char}\, K)$. Equivalently K is perfect if every algebraic extension field of K is separable over K

(see e.g., ZARISKI & SAMUEL [1], Ch. II, §§ 4-5).

8.7 <u>Corollary</u>. Let K <u>be a field. If</u> K <u>is perfect</u>, $B(K) \to B(K[X])$ <u>is an isomorphism</u>.

<u>Remark</u>. The converse is true: if K is not perfect one can explicitly give nontrivial elements of $B'(K[X])$. See AUSLANDER & GOLDMAN [1], p. 390.

<u>Proof</u>. We show $B(K) \to B(K[X])$ is surjective. Thus let A be a central separable $K[X]$-algebra; by 8.6 it suffices to find a finite extension field L of K such that $L[X]$ splits A. Let $\bar{K}$ be an algebraic closure of K. By 8.3, $B(\bar{K}(X)) = 0$. Since $\bar{K}[X]$ is a regular domain, it follows from 6.19 that $B(\bar{K}[X]) = 0$, hence A is split by $\bar{K}[X]$. But it is easy to see that if this is so, then A is already split by $L[X]$ for some finite extension L of K : given an isomorphism $\bar{K}[X] \otimes_{K[X]} A \simeq (\bar{K}[X])_n$, one captures all the matrices e_{ij} (= 1 in i-j position, 0's elsewhere) already in $L[X] \otimes_{K[X]} A$ for some finite extension L, and then clearly $L[X] \otimes_{K[X]} A \simeq (L[X])_n$. Done.

8.8 <u>Corollary</u>. Let R <u>be a regular domain of characteristic zero</u>. Then $B(R) \to B(R[X])$ <u>is an isomorphism</u>.

<u>Proof</u>. Let K be the quotient field of R. It suffices to show that the top arrow in the commutative diagram

$$\begin{array}{ccc} B(R[X]) & \longrightarrow & B(R) \\ \gamma \downarrow & & \downarrow \\ B(K[X]) & \underset{\beta}{\longrightarrow} & B(K) \end{array}$$

is injective. We therefore show instead that the arrows labelled
β and γ are injective. For β this follows from 8.7 . For
γ we need the fact that R[X] is also regular (this is nontrivial;
see e.g. NORTHCOTT [1]). Hence B(R[X]) → B(K(X)) is injective by
6.19, and therefore so is γ .

8.8 shows for example that $B(\mathbb{Z}[X_1,\ldots,X_n]) = 0$ for all n ,
since we saw in chapter 6 that $B(\mathbb{Z}) = 0$.

The reader who wonders why we ask for characteristic zero in
8.8 (instead of just perfectness of K) is referred to AUSLANDER &
GOLDMAN [2], Remark following 7.7 .

For non-perfect fields $B'(K[X])$ is nontrivial but one still
has some information:

8.9 <u>Proposition</u>. <u>Let</u> K <u>be a field of characteristic</u> p > 0 .
<u>Then</u>:

(a) <u>For any finite separable extension field</u> L <u>of</u> K , <u>the
restriction of</u> B(K[X]) → B(L[X]) <u>to</u> $B'(K[X])$ <u>has image in</u>
$B'(L[X])$ <u>and is injective</u>.

(b) <u>Every element of</u> $B'(K[X])$ <u>has order a power of</u> p .

<u>Proof</u>: (a) 8.6 guarantees that L[X] splits no nontrivial
element of $B'(K[X])$, so the injectivity on $B'(K[X])$ is clear.
From the commutative square

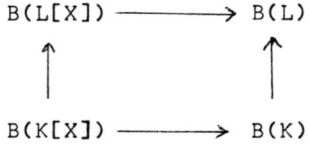

it is clear that the images of $B'(K[X]) = \ker(B(K[X]) \to B(K))$ is contained in $B'(L[X]) = \ker(B(L[X]) \to B(L))$, so (a) is proved.

(b) Let x be a nontrivial element of $B'(K[X])$. As in the proof of 8.7, x is split by $\tilde{L}[X]$ for some finite extension $\tilde{L}$ of K. (In 8.7 K was perfect and we could conclude $\tilde{L}$ is separable over K, but here we must proceed differently.) Let L be the maximal separable extension of K in $\tilde{L}$. Since $f : B(K[X]) \to B(L[X])$ is injective on $B'(K[X])$ by part (a), $y = f(x)$ has the same order as x; since $L[X]$ is a regular domain, $g : B(L[X]) \to B(L(X))$ is injective, so $y' = g(y)$ has the same order as x too. But $\tilde{L}(X)$ splits y', because $\tilde{L}[X]$ splits y, and by 3.3 and the remarks preceding 3.20 we can conclude that the order of y' divides $[\tilde{L}(X):L(X)]$. But $\tilde{L}$ is purely inseparable over L, from which it follows that $[\tilde{L}(X):L(X)] = [\tilde{L}:L]$ is a power of p; done.

CHAPTER 9. CANCELLATION.

The material of this chapter is taken from OJANGUREN & SRIDHARAN [1].

In this chapter we ask: given central separable R-algebras A, X, and Y, and an R-algebra isomorphism $A \otimes_R X \simeq A \otimes_R Y$, when can we conclude that $X \simeq Y$? If A has this property for all X and Y we say A has the <u>cancellation property</u>. A related question is: For which R do <u>all</u> central separable R-algebras have the cancellation property? These questions are admittedly artificial in the Brauer group context, since $A \otimes X \simeq A \otimes Y$ certainly implies $[X] = [Y]$ in $B(R)$. We include the results despite this, because they are interesting in themselves and because the proofs involve novel applications of several results proved in prior chapters. At the same time, since the material is not central to our theme, we have not been overly shy about quoting results from the outside world without proof when needed, so that this chapter is less self-contained than the others.

For a fixed central separable R-algebra A, we consider the following properties (which A may or may not have):

P1. For all $n \geq 1$ and all left A-modules P, $P^n \simeq A^n$ implies $P \simeq A$.

P2. For all $n \geq 1$ and all central separable R-algebras Y, $(Y)_n \simeq (A)_n$ implies $Y \simeq A$.

P3. For all central separable R-algebras X and Y, $X \otimes Y \simeq X \otimes A$ implies $Y \simeq A$.

9.1 <u>Proposition</u>. P1 <u>implies</u> P2 <u>and</u> P2 <u>implies</u> P3. <u>If every central separable R-algebra A satisfies P1 then every central separable R-algebra has the cancellation property.</u>

Remark. In fact P1, P2 and P3 are equivalent, if the conclusion in P1 is weakened from "$P \simeq A$" to "P admits a right A-structure making it a bimodule". See OJANGUREN & SRIDHARAN [1].

Proof. If we show $P1 \Rightarrow P2 \Rightarrow P3$, the rest clearly follows. Thus, assume P1 and choose an isomorphism $h : (Y)_n \to (A)_n$. We have $Y \subseteq (Y)_n$ and $A \subseteq (A)_n$ on the diagonal as usual, and we identify $(A)_n$ with $\text{End}_A(F)$ where F is the free left A-module of rank n and elements of $(A)_n$ act by right multiplication on F. Let e_{ij} be the matrix over Y with 1 in the i,j spot and zeroes elsewhere, and put $f_{ij} = h(e_{ij})$. The f's satisfy the same relations as the e's (because h is an R-algebra map), for example $f_{ij} f_{k\ell} = \delta_{jk} f_{i\ell}$. Also, since the f_{ii} are orthogonal idempotents summing to 1, we have $F = F_1 \oplus \ldots \oplus F_n$ where $F_i = F f_{ii}$. Right multiplication by f_{ji} provides an A-linear map $F_j \to F_i$ (note that for any x in F, $x f_{ji} = x f_{ji} f_{ii}$ is in F_i!), which is an isomorphism since f_{ij} provides an inverse. Thus $A^n = F \simeq F_1^n$ and therefore by P1 we have $\text{End}_A(F_1) = A \subseteq (A)_n = \text{End}_A(F)$.

We can now describe isomorphisms $Y \simeq A$. Given a in A, put $\bar{a} = \sum_{k=1}^{n} f_{k1} a f_{1k}$ in $(A)_n$, and let $\theta(a) = h^{-1}(\bar{a})$ in $(Y)_n$. Since $\bar{a} f_{ij} = f_{i1} a f_{1j} = f_{ij} \bar{a}$ we have $\theta(a) e_{ij} = e_{ij} \theta(a)$ (for all i and j), which implies that $\theta(a)$ is in Y (exercise). Hence θ is a map from A to Y; it is clearly the identity on R, and we claim it is in fact an R-algebra homomorphism. For this we must show that $\overline{ab} = \bar{a}\bar{b}$ in $\text{End}_A(F)$ (for all a,b in A). But, for each x in A, the endomorphism $\bar{x}$ leaves each summand F_i invariant, and for any pair i,j of indices, the diagram

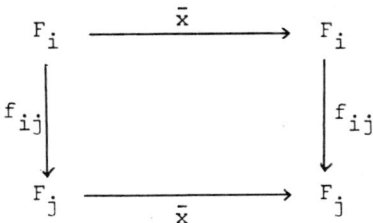

commutes. The upshot of all this is that it suffices to check $\overline{ab} = \overline{a}\overline{b}$ on $F_1 = Ff_{11}$, which is easy: since a is in $\text{End}_A(Ff_{11})$, $a = af_{11}$ on F_1, so that $xf_{11}\overline{ab} = xf_{11}abf_{11} = xf_{11}af_{11}bf_{11} = xf_{11}\overline{a}\overline{b}$.

It's clear, too, how to go back: define $\mu : Y \to \text{End}_A(F_1)$ by $\mu(y) = h(y)|F_1$. (Because y commutes with the e_{ij}, $h(y)$ commutes with f_{11}; hence F_1 is invariant under $h(y)$ and the definition is legitimate.) The verifications that θ and μ are mutual inverses are routine, and boil down to the remarks that $ye_{ij} = e_{ij}y$ and $a = af_{11}$ in $A = \text{End}_A(F_1)$ for y in Y and a in A. This completes the proof that P1 implies P2.

Finally, assume P2 and suppose $X \otimes Y \simeq X \otimes A$. Tensor with X^o to get $\text{End}_R(P) \otimes Y \simeq \text{End}_R(P) \otimes A$, P a faithfully projective R-module. Using the fact that there exist Q and $n \geq 1$ with $P \otimes Q \simeq R^n$ (see BASS [1], I.6.1) we can then tensor with $\text{End}_R(Q)$ and use 1.5(b) to get $(Y)_n \simeq (A)_n$, hence $Y \simeq A$ by P2; q.e.d.

9.2 <u>Corollary</u>. <u>If</u> R <u>is semilocal</u>, <u>every central separable</u> <u>R-algebra has the cancellation property</u>.

<u>Proof</u>. Let A be a central separable R-algebra and let P be a left A-module such that $P^n \simeq A^n$, $n \geq 1$; we show $P \simeq A$. It suffices by 1.11 to show $P/JP \simeq A/J$ where $J = \text{rad}(A)$. Using the fact (1.10) that $\mathcal{M}A \subseteq J$ where $\mathcal{M} = \text{rad}(R)$, it is clear that A/J is Artinian: it is a quotient of $A/\mathcal{M}A$, which is

finitely generated as a module over $R/\mathcal{N}$, which in turn is a finite product of fields. The fact that A/J is Artinian makes the Krull-Schmidt theorem available (see JACOBSON [3]) which in turn renders the proof trivial: $P^n \simeq A^n$ implies $(P/JP)^n \simeq (A/J)^n$ (apply $A/J \otimes_A \cdot$) , and with Krull-Schmidt this implies $P/JP \simeq A/J$.

9.3 <u>Corollary</u>. <u>If</u> R <u>is semilocal and</u> A <u>and</u> B <u>are central separable</u> R-<u>algebras with</u> [A] = [B] <u>and</u> [A:R] = [B:R] , <u>then</u> $A \simeq B$.

<u>Remark</u>. Of course, A and B need not be free; [A:R] = [B:R] means $[A_p:R_p] = [B_p:R_p]$ for all prime ideals p of R . 9.3 should be compared with the fact, noted in chapter 3, that the division-ring component of a central separable algebra over a field is unique (up to isomorphism).

<u>Proof</u>. Since [A] = [B] we have $A \otimes \text{End}_R(P) \simeq B \otimes \text{End}_R(Q)$ for faithfully projective R-modules P and Q . It suffices to show $\text{End}_R(P) \simeq \text{End}_R(Q)$. In fact, it is clear that [A:R] = [B:R] implies [P:R] = [Q:R] , and it follows that $P \simeq Q$: any two finitely generated projective modules of the same rank (at each prime) over a semilocal ring R are isomorphic. The proof of this is an easy exercise along by-now-familiar lines: it suffices to show isomorphism mod rad R by 1.11, and R/rad R is a finite product of fields; cf. 1.13(b) .

The next Corollary of 9.1 needs a lemma:

9.4 <u>Lemma</u>. <u>If</u> D <u>is a division ring, every projective</u> D[X]-<u>module is free</u>.

Proof. If D is commutative (i.e., a field) this is familiar: D[X] has the Euclidean algorithm and hence is a PID. In fact, exactly the same proof goes through in general. First, show that given f and g in D[X] with g ≠ 0, there exist q and r in D[X] with f = qg + r and deg r < deg g (exactly as in the commutative case; see LANG [1], Ch. V, §4). Then use this (as in the same reference) to show that every left ideal of D[X] is principal. A classical argument due to Kaplansky (see CARTAN & EILENBERG [1], Ch. I, §5) now shows that every projective left D[X]-module is free.

9.5 Corollary. Let R = K[X] where K is a perfect field. Then every central separable R-algebra has the cancellation property.

Proof. If A is a central separable R-algebra, 8.7 shows that [A] = [D[X]] for some central K-division algebra D. Hence by 2.19 there is a category R-equivalence

$$A - \text{Mod} \xrightleftharpoons[\psi]{\phi} D[X] - \text{Mod} \quad .$$

Now if P is a left A-module satisfying $P^n \simeq A^n$ then $(\phi P)^n \simeq (\phi A)^n$, and then $\phi P \simeq \phi A$, using 9.4. But then $P \simeq \psi\phi P \simeq \psi\phi A \simeq A$, so that A satisfies P1, and we are done by 9.1.

9.6 Corollary. Let R = K[X,Y] where K is a field of characteristic 0. Then every central separable R-algebra A has the cancellation property if and only if B(K) = 0.

"Proof". We will describe why the condition B(K) = 0 is sufficient, and only indicate briefly how necessity is proved; for

details see OJANGUREN & SRIDHARAN [1] . Assume, then, that
$B(K) = 0$. It then follows from 8.8 that $B(R) = 0$. Therefore,
for any central separable R-algebra A we have that A - Mod and
K[X,Y] - Mod are R-equivalent, by 2.19 . Now a famous theorem of
Seshadri says that projectives over R = K[X,Y] are free (see
BASS [2], Ch. IV, §6) . Hence for any R-module X , $X^n \simeq R^n$
implies $X \simeq R$. As in the proof of 9.5 , it follows that A
satisfies P1, and consequently by 9.1 that every A has the cancellation property.

Conversely, assuming $B(K) \neq 0$, let D be a non-trivial
central K-division algebra. Ojanguren and Sridharan show first
that A = D[X,Y] has a non-free right ideal P such that
$A \oplus P \simeq A \oplus A$ (i.e., Seshadri's theorem is equivalent to commutativity of the ground-field!). It follows that $P^3 \oplus A^3 \simeq A^6$,
and consequently, by Bass' cancellation theorem (see BASS [2], Ch.
IV, Cor. 3.5) that $P^3 \simeq A^3$. Hence $End_A(P^3) = (End_A(P))_3$ and
$End_A(A^3) = (A)_3$ are isomorphic R-algebras. If every central
separable K[X,Y]-algebra had the cancellation property, it would
follow (because P3 implies P2, which we have not proved here) that
$End_A(P) \simeq A$. But then P represents an element of Pic(A) ,
which is the trivial group because Pic(K[X,Y]) = 0 and A is
central separable over K[X,Y] . (See BASS [1], especially Ch. II,
§7 and Ch. III, Cor. 4.5 for a discussion of Pic(A).) This
contradicts non-free-ness of P . Done!

Remark. If K is a regular domain (e.g., a field) of characteristic 0 with $B(K) = 0$, and $N \geq 1$ is such that every projective R-module is free where $R = K[X_1,...,X_N]$, the arguments
used in 9.5 and 9.6 show that every central separable R-algebra has
the cancellation property. No pairs K, N (where K is a field
and $N \geq 1$) are known for which $K[X_1,...,X_N]$ has non-free

projective modules; various partial results, asserting that certain projective $K[X_1,\ldots,X_N]$-modules are free for certain K and/or certain N, have been proved. (The general question is known as Serre's problem.) For a survey of known results, see BASS [3].

CHAPTER 10. FAITHFULLY FLAT DESCENT.

This chapter develops some machinery for use in chapter 12. The reader familiar with descent — specifically, 10.7 and 10.9 below — can skip to chapter 11.

Throughout this chapter S is a commutative R-algebra and $\otimes$ means $\otimes_R$. If M is an R-module, M_S denotes the S-module obtained by extending scalars; it is canonically isomorphic to both $M \otimes S$ and $S \otimes M$.

We are interested in the following question: given an S-module N, is there an R-module M with $N \simeq M_S$? More precisely, what conditions must we impose to guarantee existence, and perhaps uniqueness (up to isomorphism), of such M? This question, of whether N "descends", has a satisfactory answer when S is faithfully flat over R.

10.1 Definitions. (a) An R-module M is flat if $M \otimes \cdot$ is exact, i.e., if $M \otimes \cdot$ preserves exactness.

(b) An R-module M is faithfully flat if $M \otimes \cdot$ preserves and reflects exactness.

(c) An R-algebra is flat, or faithfully flat, if it is so as R-module.

10.2 Lemmas, Examples, Exercises. (a) If T is a multiplicative set in R, $T^{-1}R$ is a flat R-algebra. As usual we write R_p for $T^{-1}R$ when T is the complement of a prime ideal p, and R_f for $T^{-1}R$ when $T = \{f^n\}_{n \geq 0}$.

(b) $\prod_{i=1}^{n} R_{f_i}$ is a faithfully flat R-algebra if $f_1, \ldots, f_n$ are such that the ideal $(f_1, \ldots, f_n)$ they generate is all of R. A direct proof, using the standard description of elements of M_f, is an easy exercise. The key fact is that $(f_1^{m_1}, \ldots, f_n^{m_n}) = R$ for any set of non-negative integers $m_1,\ldots,m_n$.

(c) *Any projective module is flat*; *any faithfully projective module is faithfully flat*. (But see BOURBAKI [2], ch. I §3, ex. 2, for an example of a faithful projective module which is not faithfully flat.)

(d) *If R is a Noetherian local ring, its completion $\hat{R}$* (5.8(c)) *is a faithfully flat R-algebra*. (See BOURBAKI [2], ch. III §3, no. 3 Prop. 6 and no. 5 Prop. 9).

(e) Show that the *$\mathbb{Z}$-algebra $\mathbb{Q}$ is faithful and flat but not faithfully flat*. (Note that the flatness follows from (a).)

(f) *Let K be a field*; show that $K[X]$ and $K(X)$ are *faithfully flat K-algebras, but $K(X)$ is not faithfully flat over $K[X]$*.

(g) Generalize (e) and (f) as follows: *let D be a domain, K its quotient field; then K is a faithfully flat D-algebra $\iff K = D$*.

(h) *Let S be a faithfully flat R-algebra, A an R-algebra. Show that A is central separable over $R \iff A_S$ is central separable over S*. (Use 2.14 and 10.4 below.)

10.3 Lemma. *Let M be an R-module, S a faithfully flat R-algebra. Then*: (a) *the canonical map $M \to M_S$ is injective, and*
(b) *S is faithful over R*, i.e., $\text{Ann}_R(S) = 0$.

Proof. (a) It's enough to show that $M \otimes S \to M \otimes S \otimes S$, $m \otimes s \mapsto (m \otimes 1) \otimes s$, is injective. But $m \otimes s \otimes t \mapsto m \otimes st$ gives a map back which splits $M \otimes S \to M \otimes S \otimes S$; done. For (b), put $M = R$ in (a), and note that the canonical map $R \to R_S \simeq S$ is the algebra structure.

Exercise. Generalize 10.3(b) as follows: if M is a faithfully flat R-module, M is faithful (i.e., $\text{Ann}_R(M) = 0$).

10.4 **Proposition.** *Let* M *be an* R-*module,* S *a faithfully flat* R-*algebra. Then:*

(a) M *is finitely generated over* $R \Leftrightarrow M_S$ *is finitely generated over* S.

(b) *Same*, *with "projective" in place of "finitely generated"*.

(c) *Same*, *with "faithfully projective" in place of "finitely generated"*.

Proof. In all three parts, " $\Rightarrow$ " is trivial. And (c) follows from (a) and (b) because of 10.3(b). The two remaining implications are left as exercises; see BOURBAKI [2], ch. I §3 no. 6, Prop. 11 and 12.

We turn now to the theorem about faithfully flat descent. The canonical reference for this is Grothendieck's S.G.A. I, exp. VIII. The more elementary presentation followed here comes from mimeographed lecture notes by M. Artin (M.I.T., 1966).

Let $S_{(n)}$ denote $S \otimes S \otimes \ldots \otimes S$ (n copies, $n \geq 1$). We have maps $d_i : S_{(n)} \to S_{(n+1)}$ ($1 \leq i \leq n+1$) and $\varepsilon_j : S_{(n+1)} \to S_{(n)}$ ($1 \leq j \leq n$) described as follows: d_i inserts 1 in the ith position, ε_j multiplies the jth and (j+1)th entries. (The notation is abusive, since for example the same symbol d_i denotes a map $S_{(n)} \to S_{(n+1)}$ for each n, but this causes less confusion than would be aroused by introducing more sophisticated subscripting.) For example, $d_2(s_1 \otimes s_2 \otimes s_3) = s_1 \otimes 1 \otimes s_2 \otimes s_3$; $\varepsilon_2(s_1 \otimes s_2 \otimes s_3 \otimes s_4) = s_1 \otimes s_2 s_3 \otimes s_4$. The d_i and ε_j are R-algebra maps, and they satisfy such identities as $\varepsilon_1 d_1 = \varepsilon_1 d_2 = \varepsilon_2 d_2 = 1$, $d_1 \varepsilon_1 = \varepsilon_2 d_1$, etc. If M is an R-module we can extend scalars to get $S_{(n)}$-modules $M_{(n)} = M_{S_{(n)}}$ and S-homomorphisms $d_i : M_{(n)} \to M_{(n+1)}$

($1 \leq i \leq n+1$) and $\varepsilon_j : M_{(n+1)} \to M_{(n)}$ ($1 \leq j \leq n$) satisfying the same identities.

The key observation is 10.6 below.

10.5 Definition. A diagram $X \to Y \underset{v}{\overset{u}{\rightrightarrows}} Z$ of abelian groups is <u>exact</u> iff $0 \to X \to Y \xrightarrow{u-v} Z$ is exact, i.e., if $X = \{y \text{ in } Y \mid uy = vy\}$ is the kernel of the pair (u,v).

10.6 Proposition. <u>Let</u> M <u>be an</u> R-<u>module</u>, S <u>a faithfully flat</u> R-<u>algebra, then</u>

$$M \to M_S \underset{d_2}{\overset{d_1}{\rightrightarrows}} M_{S \otimes S} \qquad \text{<u>is exact</u>.}$$

<u>Proof.</u> $M \to M_S$ is injective by 10.3(a), and $d_1|M = d_2|M$ is clear. For the converse we can first tensor with S to get

$$M_S \xrightarrow{d_1} M_{S \otimes S} \underset{d_2}{\overset{d_1}{\rightrightarrows}} M_{S \otimes S \otimes S} \quad .$$

Let x in $M_{S \otimes S}$ be such that $d_1 x = d_2 x$; then $x = \varepsilon_2 d_2 x = \varepsilon_2 d_1 x = d_1(\varepsilon_1 x)$, and we are done.

Now let S_i denote $S \otimes S$ viewed as S-algebra via $d_i : S \to S \otimes S$, $i = 1,2$. Let N be an S-module. Then $(N)_{S_1}$ and $(N)_{S_2}$ are $S \otimes S$-modules. To form $(N)_{S_1}$ we make $S \otimes S$ an S-algebra on the second factor, so that $(N)_{S_1} = (S \otimes S) \otimes_S N = S \otimes N$, and similarly $(N)_{S_2} = N \otimes S$; once $(N)_{S_1}$ (resp. $(N)_{S_2}$) is identified with $S \otimes N$ (resp. $N \otimes S$) in this way the $S \otimes S$-action is what one expects: $(s \otimes t)(x \otimes y) = sx \otimes ty$. Of course $S \otimes N$ and $N \otimes S$ are additively isomorphic, but <u>they are not isomorphic</u> $S \otimes S$-<u>modules in general</u> (see 10.10(b) for a concrete example). However <u>if</u>

$N \simeq M_S$ for some R-module M, then $N \otimes S$ and $S \otimes N$ are isomorphic $S \otimes S$-modules, because S_1 and S_2 are the same R-algebra. Hence a necessary condition to descend N is existence of an $S \otimes S$-isomorphism $g : N \otimes S \to S \otimes N$.

Now, given such a g, let g_i be the map derived from g by tensoring with 1_S in the ith position, $i = 1,2,3$. Thus $g_1 : S \otimes N \otimes S \to S \otimes S \otimes N$ is given by $t \otimes x \otimes s \mapsto t \otimes g(x \otimes s)$, and similarly for g_3. To write g_2 explicitly, suppose $g(x \otimes s) = \Sigma s_i \otimes x_i$; then $g_2 : N \otimes S \otimes S \to S \otimes S \otimes N$ is given by $x \otimes t \otimes s \mapsto \Sigma s_i \otimes t \otimes x_i = (1 \otimes t \otimes 1) d_2 g(x \otimes s)$. Note that g_2 and $g_1 g_3$ both go from $N \otimes S \otimes S$ to $S \otimes S \otimes N$. Moreover, if $N = M_S$ for some R-module M, then $g_2 = g_1 g_3$ (exercise). If S is faithfully flat, the converse is true:

10.7 Theorem. Let N be an S-module, S a faithfully flat R-algebra, and assume given an isomorphism $g : N \otimes S \to S \otimes N$ of $S \otimes S$-modules such that $g_2 = g_1 g_3$. Then:

(a) There is an R-module M and an S-isomorphism $\phi : M_S \to N$ such that the diagram (of $S \otimes S$-modules)

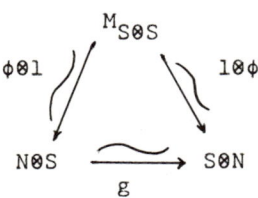

commutes, and

(b) If M', ϕ' have the properties of M, ϕ in (a), there is a unique isomorphism $M' \xrightarrow{\delta} M$ of R-modules such that $\phi \delta_S = \phi'$.

(Part (a) says that N descends, and part (b) specifies the sense in which the descended module M is unique.)

Proof. Consider the (noncommutative!) diagram of R-modules

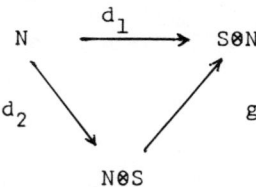

Let M be the kernel, so that

$$M \to N \underset{gd_2}{\overset{d_1}{\rightrightarrows}} S \otimes N$$

is exact. Then M is an R-module and the inclusion $M \to N$ of R-modules induces an S-homomorphism $\phi : M_S \to N$. (This is because $\operatorname{Hom}_R(M, |N|)$ and $\operatorname{Hom}_S(M_S, N)$ are naturally isomorphic for any R-module M and S-module N, where $|N|$ means N viewed as R-module — i.e., the functor "forget S" is a right adjoint for $S \otimes \cdot$.) We claim ϕ is bijective. To prove this we can first tensor with S to get the exact sequence

$$M \otimes S \to N \otimes S \underset{gd_2 \otimes 1}{\overset{d_1 \otimes 1}{\rightrightarrows}} S \otimes N \otimes S$$

of S-modules, where S acts always on the right. On the other hand, viewing N as an R-module we have that

$$N \overset{d_1}{\to} S \otimes N \underset{d_2}{\overset{d_1}{\rightrightarrows}} S \otimes S \otimes N$$

is exact (by 10.6), and if we let S act <u>on the right</u> this is a sequence of S-homomorphisms. To compare these two exact sequences, consider the square

$$
\begin{array}{ccc}
N \otimes S & \xrightarrow[gd_2 \otimes 1]{d_1 \otimes 1} & S \otimes N \otimes S \\
g \downarrow & & \downarrow g_1 \\
S \otimes N & \xrightarrow[d_2]{d_1} & S \otimes S \otimes N
\end{array}
$$

This commutes if we take the top horizontal arrows: $g_1(d_1 \otimes 1)(x \otimes s) = g_1(1 \otimes x \otimes s) = 1 \otimes g(x \otimes s) = d_1 g(x \otimes s)$. It also commutes if we take the bottom horizontal arrows:

$g_1(gd_2 \otimes 1)(x \otimes s) = g_1(g_3 d_2(x \otimes s)) = g_2 d_2(x \otimes s) = g_2(x \otimes 1 \otimes s) = d_2 g(x \otimes s)$. Consequently the vertical arrows g and g_1, which are isomorphisms, induce an isomorphism of the kernels, $M \otimes S \to N$, by the five lemma. For any x in M, $g(x \otimes 1) = gd_2 x = d_1 x = 1 \otimes x$ (the middle equality because M is the kernel of the pair d_1, gd_2). This shows that the induced isomorphism $M \otimes S \to N$ corresponds to the inclusion of M in N, i.e., <u>it is</u> ϕ; this completes the proof that $\phi : M_S \to N$ is an isomorphism.

Now looking at the kernels of the preceding diagram we have a commutative square

$$
\begin{array}{ccc}
M_S & \hookrightarrow & N \otimes S \\
\phi \downarrow & & \downarrow g \\
N & \hookrightarrow & S \otimes N
\end{array}
$$

of S-modules (with S acting on the right). The right-hand vertical is an $S \otimes S$-morphism, so we can tensor the left-hand vertical on the left by S to get a commutative diagram of $S \otimes S$-modules:

$$
\begin{array}{ccc}
M_{S \otimes S} & \longrightarrow & N \otimes S \\
1 \otimes \phi \downarrow & & \downarrow g \\
S \otimes N & \xrightarrow{id} & S \otimes N
\end{array}
$$

.

The top arrow was induced originally by the inclusion of M in N, so it is $\phi \otimes 1$. This completes the proof of (a).

For (b), suppose that M', ϕ' is also a solution to (a), and consider $\varepsilon = \phi^{-1}\phi' : M'_S \to M_S$. The exact diagram $R \to S \underset{d_2}{\overset{d_1}{\rightrightarrows}} S \otimes S$ induces a diagram

$$(**) \quad \text{Hom}_R(X,Y) \xrightarrow{\alpha} \text{Hom}_S(X_S, Y_S) \underset{\delta_2}{\overset{\delta_1}{\rightrightarrows}} \text{Hom}_{S \otimes S}(X_{S \otimes S}, Y_{S \otimes S})$$

for any R-modules X and Y : $\delta_i(f)$ is the map which makes

$$\begin{array}{ccc} X_S & \xrightarrow{f} & Y_S \\ d_i \downarrow & & \downarrow d_i \\ X_{S \otimes S} & \longrightarrow & Y_{S \otimes S} \end{array}$$

commute, and $\alpha(h) = h_S$ is the map which makes

$$\begin{array}{ccc} X & \xrightarrow{h} & Y \\ \downarrow & & \downarrow \\ X_S & \xrightarrow{\alpha(h)} & Y_S \end{array}$$

commute. <u>Suppose we know that</u> $(**)$ <u>is also exact</u>, <u>and that</u> $\delta_1 \varepsilon = \delta_2 \varepsilon$ (<u>when</u> $X=M'$, $Y=M$). Then $\varepsilon = \delta_S$ for a unique δ in $\text{Hom}_R(M',M)$, which is necessarily an isomorphism, and $\delta_S = \varepsilon = \phi^{-1}\phi'$ shows that $\phi \delta_S = \phi'$ as required. Thus the proof is complete if we show:

10.8 Lemma. <u>With hypotheses and notations as above</u>, $(**)$ <u>is exact. Taking</u> $X=M'$, $Y=M$, <u>we have</u> $\delta_1 \varepsilon = \delta_2 \varepsilon$.

Proof. Any h in $\text{Hom}_R(X,Y)$ is determined by its image $\alpha(h) = h_S$ in $\text{Hom}_S(X_S, Y_S)$ because $X \to X_S$ and $Y \to Y_S$ are injective. Hence α is injective, and clearly $\delta_1 \alpha = \delta_2 \alpha$. Conversely, suppose f in $\text{Hom}_S(X_S, Y_S)$ is such that $\delta_1 f = \delta_2 f$. Viewing $X \to X_S$, $Y \to Y_S$, and $\text{Hom}_R(X,Y) \to \text{Hom}_S(X_S, Y_S)$ as inclusions, we have to show that $\text{Im}(f|X) \subseteq Y$. Since $Y = \{z \text{ in } Y_S \mid d_1 z = d_2 z\}$, what we need is $d_1 f x = d_2 f x$ for all x in X. But for x in X, $d_1 f x = (\delta_1 f) d_1 x = (\delta_2 f) d_2 x = d_2 f x$ (in the middle equality, d_1 and d_2 are maps from X_S to $X_{S \otimes S}$ and they agree on X).

Now all that's left is to check that $\delta_1 \varepsilon = \delta_2 \varepsilon$. Note first that because $\phi \varepsilon = \phi'$, both of the following diagrams commute:

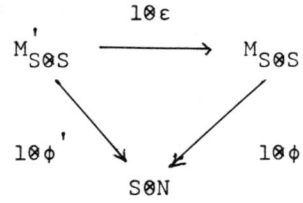

and

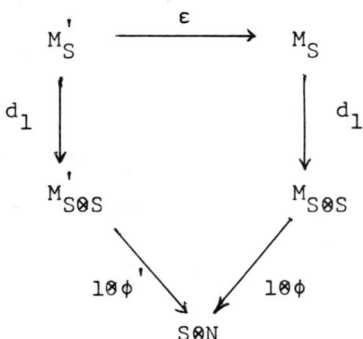

Therefore, since $\delta_1 \varepsilon$ is <u>the</u> map $M'_{S \otimes S} \to M_{S \otimes S}$ making the square in the second diagram commute, we have $\delta_1 \epsilon = 1 \otimes \epsilon$. Similarly, $\delta_2 \epsilon = \epsilon \otimes 1$. But the two resulting diagrams are joined by g :

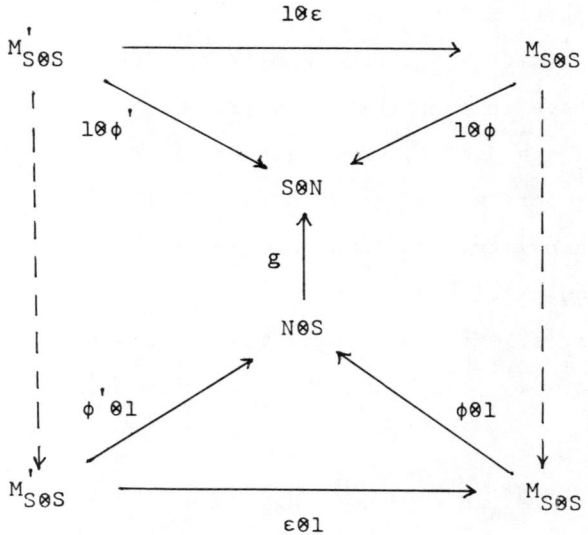

By part (a), this latter diagram commutes if the dotted arrows are the identity maps. Hence $\delta_1 \epsilon = 1 \otimes \epsilon = \epsilon \otimes 1 = \delta_2 \epsilon$, and the proof is complete.

Finally, suppose N is an S-algebra; then $N \otimes S$ and $S \otimes N$ are $S \otimes S$-algebras, and we have:

10.9 Afterthought. <u>In the setting of theorem 10.7, assume that N is an S-algebra and the descent map $g : N \otimes S \to S \otimes N$ is an $S \otimes S$-algebra isomorphism. Then there is a unique R-algebra structure on M making $\phi : M_S \to N$ an S-algebra isomorphism.</u>

<u>Proof</u>. The multiplication on N is an S-homomorphism $N \otimes_S N \to N$; using ϕ to pull it back, we have a map $M_S \otimes_S M_S \to M_S$, i.e., an element μ of $\text{Hom}_S((M \otimes M)_S, M_S)$. Viewing $S \otimes S$ as an

S-algebra in two ways (via d_1 and d_2) as before, we get two multiplications on $M_{S \otimes S}$, viz. $\delta_1 \mu$ and $\delta_2 \mu$ in $\text{Hom}_{S \otimes S}((M \otimes M)_{S \otimes S}, M_{S \otimes S})$. Explicitly, $\delta_1 \mu$ is the multiplication on $M_{S \otimes S}$ pulled back from that of $N \otimes S$ via the $S \otimes S$-module isomorphism $\phi \otimes 1$, and $\delta_2 \mu$ is pulled back from $S \otimes N$ via $1 \otimes \phi$. Since

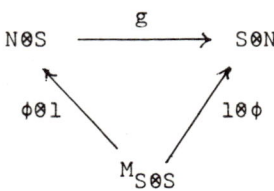

commutes, and g is an $S \otimes S$-<u>algebra</u> isomorphism, the two multiplications $\delta_1 \mu$ and $\delta_2 \mu$ coincide. Thus, by exactness of (**) (with $X = M \otimes M$, $Y = M$), μ comes from a unique element of $\text{Hom}_R(M \otimes M, M)$; this just says there is a unique R-multiplication on M carried by $\phi : M_S \to N$ to the S-multiplication we started with. Associativity of the multiplication on N is the assertion that two maps $(M \otimes M \otimes M)_S = N \otimes N \otimes N \to N = M_S$ are equal, i.e., their difference in $\text{Hom}_S((M \otimes M \otimes M)_S, M_S)$ is zero. Again by exactness of (**), the corresponding map in $\text{Hom}_R(M \otimes M \otimes M, M)$ is zero, i.e., the multiplication on M is associative. (In exactly the same way one can show that if N has any property expressible in a reasonable way in terms of the agreement of two maps, for example commutativity, then M inherits the property.) Finally, the multiplication on M has a unit 1_M: the diagram

$$M \to M_S = N \underset{d_2}{\overset{d_1}{\rightrightarrows}} M_{S \otimes S}$$

is exact, and $d_1(1_N) = d_2(1_N)$ because both are unit elements in $M_{S \otimes S}$, etc.

10.10 **Exercise**. (a) Let N be an S-module; convince yourself that the obvious map $N \otimes S \to S \otimes N$ can fail to be an $S \otimes S$-isomorphism.

(b) Let $S = R[X]$, R a field, and $N = S/(X)$. Show that $N \neq M_S$ if $M = (N$ viewed as R-module). Show that $N \otimes S$ and $S \otimes N$ are not isomorphic $S \otimes S$-modules. (This example is due to M. Artin.)

CHAPTER 11. SPLITTING RINGS OVER LOCAL RINGS.

The goal of this chapter is 11.3 , which says that central separable algebras over local rings admit good splitting rings. This will be an important ingredient of the proof, in chapter 12 , that $B(R)$ is always a torsion group.

11.1 <u>Lemma</u>. <u>Let</u> $A = (F)_n$, F <u>any field</u>, <u>and let</u> W <u>be a commutative subalgebra of</u> A . <u>Then</u>:

(a) $W \subseteq A^W$, <u>and</u> $W = A^W \Leftrightarrow W$ <u>is a maximal commutative subalgebra</u>.

(b) <u>If</u> $W = \{(a_{ij}) \mid a_{ij} = 0 , \text{ all } i \neq j\}$ <u>then</u> W <u>is a maximal commutative subalgebra of</u> A .

(c) <u>Let</u> W <u>be an extension field of</u> F <u>with</u> $[W : F] = n$, W <u>being viewed as a subalgebra of</u> $A \simeq \text{End}_F(W)$ <u>via multiplications</u>. <u>Then</u> W <u>is a maximal commutative subalgebra of</u> A.

Before proving this we make a few remarks. 11.1(b) and (c) provide two examples of maximal commutative subalgebras of $(F)_n$ of dimension n . It should be pointed out that maximal commutative subalgebras of $(F)_n$ need not have dimension n . For example* the matrices $xI + \begin{pmatrix} 0 & A \\ 0 & 0 \end{pmatrix}$, where A is 2×2 , form a commutative subalgebra of dimension 5 in $(F)_4$; the matrices $xI + \begin{pmatrix} 0 & B \\ 0 & 0 \end{pmatrix}$ where B is 2×3 form a commutative subalgebra of dimension 7 in $(F)_5$, and in the same way one has commutative subalgebras of $(F)_n$ of dimension $[\frac{n^2}{4}] + 1$ for any n . (If $n = 2m+1$ is odd, $[\frac{n^2}{4}] = (m+1)m$.) A theorem of Schur and Jacobson says that no commutative subalgebra can have dimension greater than $[\frac{n^2}{4}] + 1$ (see SUPRUNENKO & TYSHKEVICH [1], p.95) . On the other hand, it is quite possible to have maximal commutative

---

* pointed out to us by L. Roberts.

subalgebras of $(F)_n$ of dimension $< n$; see COURTER [1] . Courter displays a 13-dimensional maximal commutative subalgebra of $(F)_{14}$ (for any F), and uses it to show that the g.l.b. of the numbers r/n , r = dimension of a maximal commutative subalgebra of $(F)_n$ for some n , is zero. Courter also supplies references to the papers of Schur and Jacobson, about the upper bound $[\frac{n^2}{4}] + 1$.

Proof of 11.1. (a) is clear, and shows that in (b) and (c) we need only prove $A^W \subseteq W$. For (b), if a matrix α commutes with everything in W , then α commutes in particular with the e_{ii} (1 in i-i spot, 0 elsewhere), and this is enough to force α to be in W (exercise). For (c) it is enough to notice that $End_F(W)^W = End_W(W) = W$.

It is convenient to record here for reference:

11.2 <u>Lemma</u>. <u>Let</u> $x_1,\ldots,x_n$ <u>be a set of generators for an</u> R-<u>module</u> M . <u>If</u> M <u>is free of rank</u> n <u>then</u> $x_1,\ldots,x_n$ <u>is a basis</u>.

This is a consequence of 1.13(c), or 1.14 .

11.3 <u>Theorem</u>. <u>Let</u> A <u>be a central separable algebra over the Noetherian local ring</u> R . <u>There is a splitting ring</u> S <u>for</u> A (<u>definition</u> 3.13) <u>such that</u> S <u>is a faithful</u> R-<u>algebra</u> (i.e. R $\subseteq$ S) , S <u>is free of finite rank as</u> R-<u>module</u>, <u>and</u> S <u>is a separable</u> R-<u>algebra</u>. <u>If</u> R <u>is complete</u>, S <u>is local</u>; <u>in general</u>, S <u>is semilocal</u>.

The proof we shall give is attributed by Auslander and Goldman to Serre. The idea is to capitalize on the fact, noted in 3.15,

that everything is lovely modulo the maximal ideal $\mathfrak{m}$ of R :
over $R/\mathfrak{m}$ there is a good splitting field for $R/\mathfrak{m} \otimes_R A$. One
lifts this in the obvious way, thereby disposing of the complete
case. The general case is settled by passing to the completion
of R .

Proof of 11.3. Write $\bar{R}$ for $R/\mathfrak{m}$, $\bar{X}$ for $R/\mathfrak{m} \otimes_R X$.
Choose a finite separable field extension L of $\bar{R}$ which splits
$\bar{A}$ (see 3.15) . L is generated by a primitive element:
$L = \bar{R}(\theta)$. Let f in $\bar{R}[X]$ be the (monic, irreducible) minimal
polynomial of θ , so that $L \simeq \bar{R}[X]/(f)$. Choose F in $R[X]$
such that $\bar{F} = f$ and F is monic, with $\deg F = \deg f \geq 1$, and
put $S_1 = R[X]/(F)$. Then we claim:

(1) S_1 is a faithful R-algebra.
(2) S_1 is free of finite rank as R-module.
(3) S_1 is a separable R-algebra.
(4) S_1 is local with maximal ideal $\mathfrak{m} S_1$.
(5) L splits $S_1 \otimes A$.

Of these, (1) and (2) are clear. Since $\bar{S}_1 = S_1/\mathfrak{m} S_1 \simeq L$, (3)
follows from 4.12 and 4.7 ; and (4) follows from 4.3(a) . (5)
follows from the fact that L splits $\bar{A}$, because of the
commutative diagram

$$\begin{array}{ccc} B(S_1) & \longrightarrow & B(\bar{S}_1) = B(L) \\ \uparrow & & \uparrow \\ B(R) & \longrightarrow & B(\bar{R}) \end{array}$$

Now if R is complete, S_1 is also complete, by 5.5 .
Hence in this case (5) implies that S_1 splits A , for

$B(S_1) \to B(L)$ is injective by 5.7 . Hence when R is complete we take $S = S_1$ and we are done. Also, (5) together with 4.4 shows that for the general case we can assume $\bar{A} \sim 1$ (by working over S_1 instead of R). So for the rest of the proof we add the hypothesis that $\bar{A} = (\bar{R})_n$.

We claim that there is a maximal commutative $\bar{R}$-subalgebra W of $\bar{A}$, of dimension n over $\bar{R}$, such that W is a separable $\bar{R}$-algebra of the form $W = \bar{R}[\alpha]$, and with basis $1, \alpha, \ldots, \alpha^{n-1}$ over $\bar{R}$. Namely, take W to be the "diagonal" matrices as in 11.1(b) . W is separable because it is isomorphic to $\bar{R} \times \ldots \times \bar{R}$ (n copies), by 4.5 . To see that W is of the form $\bar{R}[\alpha]$, let α be the diagonal matrix $\mathrm{diag}(a_1, \ldots, a_n)$ where the a_i are <u>distinct</u> elements of $\bar{R}$; then the inclusion $\bar{R}[\alpha] \subseteq W$ is an equality because the minimal polynomial of α visibly has degree n , so that $\dim \bar{R}[\alpha] = n = \dim W$. The only catch in this construction is that $\bar{R}$ might not have as many as n distinct elements. But whenever $\bar{R}$ is finite we can satisfy the claim in another way: let W be a separable field extension of $\bar{R}$ of degree n , and embed it in $\bar{A} = \mathrm{End}_{\bar{R}}(W)$ as in 11.1(c) . This proves the claim.

Now choose β in A such that $\bar{\beta} = \alpha$, and let S be the R-submodule of A generated by $1, \beta, \ldots, \beta^{n-1}$. Choose $\gamma_1, \ldots, \gamma_m$ in $\bar{A}$ so that $1, \alpha, \ldots, \alpha^{n-1}, \gamma_1, \ldots, \gamma_m$ is an $\bar{R}$-basis for $\bar{A}$, and choose δ_i in A such that $\bar{\delta}_i = \gamma_i$. Then, by Nakayama's Lemma (1.9), the elements $1, \beta, \ldots, \beta^{n-1}$, $\delta_1, \ldots, \delta_m$ generate A as R-module, and then by 11.2 they are a basis. Hence S is a finitely generated free R-module with $S \supseteq R$, and S is an R-direct summand of A . <u>We claim</u> β^n <u>is in</u> S . It then follows that S is a commutative subalgebra of A , which is R-separable by 4.12 . Thus once we have β^n i S   it will remain to show only that S splits A .

To show that β^n is in S we apply $\hat{} = \hat{R} \otimes_R \cdot$ where $\hat{R}$ is the completion of R (see 5.8(c)) . Since S is a direct summand of A , $\hat{S}$ is a direct summand of $\hat{A}$; and it suffices (because $\hat{R}$ is faithfully flat over R — see 10.2(d)) to show that the image of β^n (under $A \to \hat{A}$) lands in $\hat{S}$. Now the commutative diagram

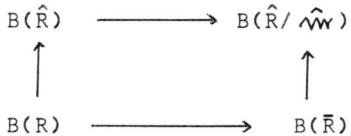

$$\begin{array}{ccc} B(\hat{R}) & \longrightarrow & B(\hat{R}/\widehat{\mathfrak{m}}) \\ \uparrow & & \uparrow \\ B(R) & \longrightarrow & B(\bar{R}) \end{array}$$

shows that $\hat{A}$ is trivial in $B(\hat{R})$, for $\bar{A}$ is trivial in $B(\bar{R})$ and the top horizontal is injective by 5.7 . Thus $\hat{A} \simeq \text{End}_{\hat{R}}(\hat{F})$ for some free $\hat{R}$-module $\hat{F}$, and $[\hat{F} : \hat{R}] = n$. Now view β as an element of $\hat{A}$, hence as an endomorphism of $\hat{F}$. Its characteristic polynomial $\phi(X)$ is monic, of degree n , with coefficients in R , and $\phi(\beta) = 0$ by Cayley-Hamilton (4.1) . Hence β^n is a linear combination, over $\hat{R}$, of 1 , $\beta,\ldots,\beta^{n-1}$. But this shows that β^n is in $\hat{S}$, hence β is in S , as claimed.

It remains to see that S splits A . <u>We claim first that</u> $A^S = S + (A^S \cap \mathfrak{m} A)$. Since $\bar{S} = W$ is a maximal commutative subalgebra of $\bar{A}$ we have $\bar{A}^{\bar{S}} = \bar{S}$; thus $A^S \subseteq \{x \text{ in } A \mid \bar{x}y = y\bar{x} \text{ for all } y \text{ in } \bar{S}\} = \{x \text{ in } A \mid \bar{x} \text{ is in } \bar{S}\} = S + \mathfrak{m} A$. The claim follows. We will see below that in fact $A^S = S$; this does not follow immediately from what we just did because $A^S \cap \mathfrak{m} A$ is not, <u>a priori</u>, contained in $\mathfrak{m} A^S$.

Viewing A as a left S-module, we know that it is projective by 2.4, hence $\text{End}_S(A)$ is a central separable S-algebra (see

2.17). Now the canonical isomorphism $\eta : A^e \to \text{End}_R(A)$ gives rise to a commutative diagram

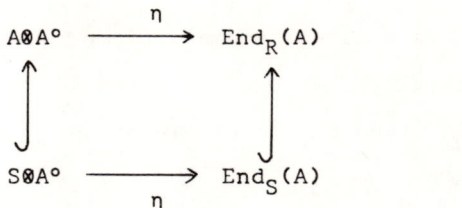

viewing the injections as inclusions, we have $A^S = A^S \otimes 1 \subseteq \text{End}_S(A)^{S \otimes A^\circ} \subseteq \text{End}_R(A)^{S \otimes A^\circ} = (A \otimes A^\circ)^{S \otimes A^\circ} = A^S \otimes 1$, so that A^S is the centralizer, in $\text{End}_S(A)$, of $S \otimes A^\circ$. By the double-centralizer theorem (2.13 and 2.15(b)) we have therefore that A^S is central separable over S, and $(S \otimes A^\circ) \otimes_S A^S \to \text{End}_S(A)$ is an isomorphism. If we now show that the inclusion $S \to A^S$ is an equality we are done.

For this we use the facts, established above, that A^S is central separable over S, and that $A^S = S + (A^S \cap \mathfrak{m} A)$. Since $A^S \cap \mathfrak{m} A$ is a two-sided ideal in A^S, there is an ideal I in S with $IA^S = A^S \cap \mathfrak{m} A$ and $IA^S \cap S = I$ (see 2.11). This gives $I = A^S \cap \mathfrak{m} A \cap S = \mathfrak{m} A \cap S = \mathfrak{m} S$, the latter because S is a direct summand of A. Hence $A^S = S + (A^S \cap \mathfrak{m} A) = S + \mathfrak{m} SA^S = S + \mathfrak{m} A^S$, and (by Nakayama) therefore $A^S = S$. This completes the proof.

<u>Remarks</u>. (1) The only point at which the assumption that R is Noetherian intervenes in the proof is where we pass from R to $\hat{R}$ to show β^n is in S. One can give a proof without the Noetherian hypothesis by invoking Henselization rather than completion. The Noetherian case will be sufficient for our applications, because of 12.1 below.

(2) If R, A, S are as in 11.3, we have an isomorphism $S \otimes_R A \to \text{End}_S(P)$ where P is faithfully projective over S. In fact, P is free. For A is free over R by 1.12, hence $\text{End}_S(P)$ is free over S. In particular, $\text{End}_S(P)$ has constant rank as an S-module, and therefore so does P. But S is semilocal, and we have seen that projectives of constant rank over semilocal rings are free (1.13(b)).

APPENDIX. (G. Garfinkel) This section is devoted to proving 11.5 below. A discussion of what it means for R to be regular was given in 6.20. It will suffice for our discussion here to recall that R is Noetherian and integrally closed, that the same is true for each R_p, p a prime, and that R_p has finite global dimension. It follows that each R_p is a UFD, a deep fact we shall assume without proof (see KAPLANSKY [2], Theorem 184, p. 135). For the sake of compactness, a number of exercises are scattered within the proof. However, the lemma below is used often, so will be isolated. Only an indication of its proof is given.

11.4 <u>Lemma</u>. Let R <u>be a Noetherian domain</u>, M, N <u>finitely generated</u> R-<u>modules</u>, S <u>a commutative</u> R-<u>algebra which is a flat</u> R-<u>module. Then</u>:

(a) <u>The natural map</u> $S \otimes_R \mathrm{Hom}_R(M,N) \to \mathrm{Hom}_S(S \otimes_R M, S \otimes_R N)$ <u>is an isomorphism</u>.

(b) <u>If</u> M <u>is</u> R-<u>reflexive, then</u> $S \otimes_R M$ <u>is</u> S-<u>reflexive</u>.

(c) <u>If</u> M_p <u>is</u> R_p-<u>projective for all primes</u> p <u>of</u> R, <u>then</u> M <u>is</u> R-<u>projective</u>.

<u>Proof</u>. (a) The key feature is that M is finitely presented, i.e., there is an exact sequence $F_2 \to F_1 \to M \to 0$ with F_i free of finite type. (See BASS [1], Cor. 2.8, p. 93 for a proof.)

(b) follows from (a).

(c) Since R is a domain, M has constant rank, say n (see 1.13). The key is to show that for each maximal ideal p, there exists f in R, not in p, with M_f projective (hence free of rank n) over R_f; see BOURBAKI [2], ch. II, §5, no. 2 . The f's arising in this way (one outside each p) generate the unit ideal, so a finite number of them, say $f_1,\ldots,f_m$, suffice. Then $R_{f_1} \times \ldots \times R_{f_m}$ is faithfully flat over R (10.2(b)) , and M is R-projective by 10.4(b) .

11.5 <u>Theorem</u>. <u>Let</u> R <u>be a regular domain</u>, M <u>a finitely generated reflexive</u> R-<u>module</u>. <u>Suppose</u> $\text{End}_R(M)$ <u>is</u> R-<u>projective</u>. <u>Then</u> M <u>is</u> R-<u>projective as well</u>.

<u>Proof</u>. Taking $S = R_p$ in 11.4 (a) and (b) , we see that the hypotheses on M remain valid after localizing. By 11.4(c) we may then assume R is local, hence a UFD . For an arbitrary R-module X , we shall write $\text{rank}_R X$ to mean $\dim_K(K \otimes_R X)$. (This agrees with the notion of rank for projectives; see 1.13(a).) We may assume $M \neq 0$. Then M is torsion free, since it is reflexive (6.13(b)) . Hence M has rank ≥ 1 (Exercise; or see CARTAN & EILENBERG [1], Prop. 2.1, p. 130). Let us first consider the case when M has rank 1 . Then M is isomorphic to an ideal of R (Exercise). For I any ideal of R , $I^* = \text{Hom}_R(I,R)$ may be identified with $I^{-1} = \{a \text{ in } K \mid aI \subseteq R\}$: for any $f : I \to R$ satisfies $f(xy) = xf(y) = f(x)y$, hence $a = f(x)x^{-1}$ $(x \neq 0)$ is in I^{-1} and independent of x . The reflexivity of M is then easily seen to imply that $(I^{-1})^{-1} = I$ (we are using the fact that the identification of I^* with I^{-1} is valid for I any finitely generated R-submodule of K ; I^* is finitely generated by 6.7). It is not hard to see that for J any finitely generated R-submodule of K , $J^{-1} = \bigcap_{0 \neq x \in J} Rx^{-1}$. Since $I = (I^{-1})^{-1}$ we get

$$I = \bigcap_{0 \neq x \in I^{-1}} Rx^{-1} \quad .$$

For x in I^{-1} write $x^{-1} = u(x)v(x)^{-1}$, with $u(x)$, $v(x)$ in R and relatively prime (recall that R is now a UFD). If $\underline{a}$ is in I, $\underline{a} = rx^{-1} = ru(x)v(x)^{-1}$, and because $\gcd(u(x), v(x)) = 1$ we conclude that $\underline{a}$ is in $Ru(x)$. Thus

$$I \subseteq \bigcap_{0 \neq x \in I^{-1}} Ru(x) \subseteq \bigcap_{0 \neq x \in I^{-1}} Rx^{-1} = I$$

and $I = \bigcap Ru$, where u ranges over a set of elements in R. It is straightforward to check that this intersection is principal when R is a UFD. Thus I is R-free, which disposes of the case where M is of rank 1. It may be worthwhile to note that the hypothesis that $\text{End}_R(M)$ is projective didn't seem to be used because $\text{End}_R(I) = R$ automatically when I is an ideal of R, the proof being essentially the same as that $I^* = I^{-1}$.

Now let M be of any rank, and let $A = \text{End}_R(M)$. A is a central separable R-algebra: if p is a minimal nonzero prime of R, M_p is a finitely generated reflexive R_p-module. Since R_p is a discrete valuation ring, hence a PID, M_p is then R_p-free. Then $\text{End}_{R_p}(M_p)$ is a central separable R_p-algebra; but $\text{End}_{R_p}(M_p) \simeq R_p \otimes_R \text{End}_R(M)$ (11.4). Hence A_p is a central separable R_p-algebra for all p of height 1. The center of $\text{End}_R(M)$ is R : for any element f of it is in K since f is in $K \otimes \text{End}_R(M) \simeq \text{End}_K(K \otimes M)$ (11.4) ; moreover f is integral over R by the Cayley-Hamilton theorem (4.1), and since R is a UFD it is integrally closed in K. By 6.30, A is a central separable R-algebra.

Let S be a splitting ring for A which is a free R-module of finite rank and a separable R-algebra (11.3). Then $S \otimes_R \text{End}_R(M) \simeq \text{End}_S(Q)$ where Q is a faithfully projective S-module. Then $\text{End}_S(S \otimes_R M) \simeq \text{End}_S(Q)$. If we could show that $S \otimes_R M$ was S-projective, it would follow that M is R-projective (10.2(c), 10.4(b)). We shall now reduce to the case where S is a regular local splitting ring of A.

Since S is separable over R, every R-projective S-module is S-projective (2.4). Since S is R-free, any S-projective S-module is R-projective. It follows that S has global dimension at most (in fact equal to) the global dimension of R: For if gl.dim $R \leq n$ then in any exact sequence $0 \to P_n \to P_{n-1} \to \ldots \to P_0 \to L \to 0$, with L an R-module and P_i R-projective for $i < n$, P_n must be R-projective; the rest is an easy exercise. Since S has finite global dimension, so does each S_P, P a prime of S (see 6.20); hence S_P is a regular local ring. S_P is still a splitting ring for A over R, being an extension of S. Suppose we could show that $S_P \otimes_S S \otimes_R M \simeq S_P \otimes_R M$ were projective for all P. The rank n_P of $S_P \otimes_R M$ over S_P would equal n, the rank of M over R, as seen by considering that $S_P \otimes_R K \otimes_R M$ is free of rank both n_P and n. It would follow from 11.4(c) that $S \otimes_R M$ is S-projective. Finally, the hypotheses that $S \otimes M$, Q, satisfy over S are preserved on passage to S_P, by 11.4 and 1.6.

Hence we may assume we have S a regular local domain, with $N = S \otimes_R M$ an S-module satisfying $\text{End}_S(N) \simeq \text{End}_S(Q)$, with Q a faithfully projective S-module.

Let $E = \text{End}_S(Q)$. The isomorphism $\text{End}_S(N) \simeq E$ allows us to view N as a left E-module. It follows that $N = Q \otimes_S L$ for some S-module L (This follows from 1.18, or can be seen directly by taking $L = Q^* \otimes_E N$ ($Q^* = \text{Hom}_S(Q,S)$) and setting up an explicit isomorphism.) It follows from the isomorphism $\text{End}_S(N) \simeq \text{End}_S(Q)$ that N and Q have the same rank over S, hence that L has rank 1. We claim that L is reflexive.

We leave as an exercise the verification that the natural maps $Q^* \otimes L^* \to (Q \otimes L)^*$, $(Q \otimes L)^{**} \to (Q^* \otimes L^*)^*$, $Q^{**} \otimes L^{**} \to (Q^* \otimes L^*)^*$ are isomorphisms for Q a finitely generated projective S-module. Let $\alpha_X : X \to X^{**}$ be the natural map, and recall that $\alpha_{Q \otimes L} : Q \otimes L \to (Q \otimes L)^{**}$ is an isomorphism because $N = Q \otimes L$ is a reflexive S-module. The commutative diagram below shows that $1_Q \otimes \alpha_L$ is an isomorphism:

$$\begin{array}{ccc}
Q \otimes L & \xrightarrow{1_Q \otimes \alpha_L} & Q \otimes L^{**} \\
{\scriptstyle \alpha_{Q \otimes L}} \searrow & & \swarrow {\scriptstyle \alpha_Q \otimes 1} \\
(Q \otimes L)^{**} & & Q^{**} \otimes L^{**} \\
& \searrow \quad \swarrow & \\
& (Q^* \otimes L^*)^* &
\end{array}.$$

Since Q is faithfully projective it is faithfully flat (10.2(c)), hence α_L is an isomorphism.

Thus L is a rank 1 reflexive S-module with S a regular local ring. The argument towards the beginning of this proof showed that in this case L is S-projective. Since $N = Q \otimes L$, N is S-projective (1.5). This concludes the proof.

CHAPTER 12. THE BRAUER GROUP IS TORSION.

We are finally ready to prove that for any commutative ring R , B(R) is a torsion group, i.e. given a central separable R-algebra A there is an isomorphism (as R-algebras) between $A \otimes_R \ldots \otimes_R A$ (n copies, for some n), and $\text{End}_R(P)$ for some faithfully projective R-module P . In fact we show that if the rank of A is constant, say $[A : R] = m^2$ (cf. remark after 3.2), then n = m works.

This was first proved about 1965 by very sophisticated means, e.g., étale cohomology of schemes; see GROTHENDIECK [1] and GIRAUD [1], V, §4 . A more accessible route to this theorem is indicated in KNUS & OJANGUREN [2], and it is their method we follow here. A different proof, for the case where R is an integrally closed Noetherian domain, can be found in B. AUSLANDER [2] . The essential ingredients of the Knus-Ojanguren proof are the technique of faithfully flat descent (chapter 10) and the existence of good splitting rings over local rings (theorem 11.3) . We are indebted to M. Knus for some correspondence clarifying some details of the proof.

We begin with an important technical lemma:

12.1 Lemma. (KNUS & OJANGUREN [1], 3.2) Let A be a central separable R-algebra. There is a Noetherian subring R' of R and a central separable R'-algebra A' contained in A , such that $R \otimes_{R'} A' = A$.

Proof. Since A is a finitely generated projective R-module, we can write A as coker(f) where f is an idempotent endomorphism of R^n , for some n . Let α be a matrix for f , say with respect to the standard basis, and let R_1 be the subring of R generated by the entries of α. Then α defines an idempotent endomorphism of R_1^n, whose cokernel A_1 is therefore a finitely generated projective R_1 module contained in A, and $R \otimes_{R_1} A_1 = A$. Let $\{x_1, \ldots, x_m\}$ be a set of generators for the R_1-module A_1, hence also for the R-module A, and define c_{ijk} and c_ℓ in R by $x_i x_j = \sum_k c_{ijk} x_k$ and $1 = \sum_\ell c_\ell \otimes a_\ell$, a_ℓ in A_1. Let R_2 be the subring of R generated by R_1, the c_{ijk}, and the c_ℓ. Then $A_2 = R_2 \otimes_{R_1} A_1$ is an R_2-algebra contained in A, and $A = R \otimes_{R_2} A_2$. Let $\eta : A \otimes_R A^\circ \to End_R(A)$ be the canonical isomorphism, and choose $y_1, \ldots, y_r$ in $A \otimes A^\circ$ such that $\eta(y_1), \ldots, \eta(y_r)$ generate the R_2-module $End_{R_2}(A_2)$ (it is finitely generated by 6.7). Each y_i can be written as a sum $\sum_{j,k} d_{ijk} x_j \otimes x_k^\circ$ with d_{ijk} in R; let R' be the subring of R generated by R_2 and the d_{ijk}. The R'-algebra $R' \otimes_{R_2} A_2$ is central separable by 2.14 since, by construction, the restriction of η to $A' \otimes_{R'} (A')^\circ$ is an isomorphism onto $End_{R'}(A')$, and A' is faithful over R' (because $A = R \otimes_{R'} A'$), and finitely generated and projective (because A_1 was so over R_1). R' is Noetherian because it is generated over the prime subring of R by finitely many elements (i.e., R' is a quotient of $\mathbb{Z}[X_1, \ldots, X_t]$ for some t).

12.2 <u>Definition</u>. Let A be an R-algebra, not necessarily central separable. If $\sigma : S \otimes_R A \to End_S(P)$ is an S-algebra isomorphism, where S is a commutative R-algebra and P is a faithfully projective S-module, we say that (S, P, σ) is a <u>splitting</u> for A.

12.3 <u>Theorem</u>. <u>Let</u> A <u>be an</u> R-<u>algebra</u>, R <u>Noetherian</u>. <u>The</u>

following are equivalent:

(1) A <u>is central separable over</u> R .

(2) <u>For each prime ideal</u> p <u>of</u> R , <u>there exists</u> f <u>in</u> R , <u>not in</u> p , <u>such that</u> A_f <u>has a splitting</u> (S(f), P(f), σ(f)) <u>with</u> S(f) <u>separable and free of finite rank over</u> R_f <u>and</u> P(f) <u>free of finite rank over</u> S(f) .

(3) A <u>has a splitting</u> (S, P, σ) <u>with</u> S <u>faithfully flat over</u> R <u>and</u> S <u>finitely generated as</u> R-<u>algebra</u>.

(4) A <u>has a splitting</u> (S, P, σ) <u>with</u> S <u>faithfully flat over</u> R .

Remark. 11.3 shows that given A central separable over R , we have a good splitting at each point p of Spec(R) . Condition (2) says that we have the same in a neighborhood U = Spec(R_f) of each point.

Proof. We show (1) ⇒ (2) ⇒ (3) and (4) ⇒ (1) .

(1)⇒(2): Given p , we have by 11.3 a separable R_p-algebra S(p) free of finite rank over R_p , a free S(p)-module P(p) of finite rank (cf. remark 2 following 11.3), and an isomorphism σ(p) : S(p)⊗$_{R_p}$$A_p$ → End$_{S(p)}$(P(p)) of S(p)-algebras. As in the proof of 12.1, there is a subring R' of R_p generated by finitely many elements, a central separable R'-algebra A' contained in A_p , a separable R'-algebra S' , free of finite rank over R' and contained in S(p) , and a free S'-module P' contained in P(p) , such that: R_p⊗$_{R'}$A' = A_p , R_p⊗$_{R'}$S' = S(p) , S(p)⊗$_{S'}$P' = P(p) , and the restriction σ' of σ(p) is an isomorphism S'⊗$_{R'}$A' → End$_{S'}$(P') of S'-algebras. Say $z_1,\ldots,z_n$ are elements of R_p which generate the subring R' . Each z_i can be written (not uniquely) in the form r_i/f_i with r_i, f_i in R and f_i not in p. Let $f = \prod_{i=1}^{n} f_i$; then f is not in p, R_f is an R'-algebra and R_p is an R_f-algebra via obvious maps, and $A_f = R_f \otimes_{R'} A'$. Finally, let $X(f) = R_f \otimes_{R'} X'$ for $X = S$, P and σ; then $(S(f), P(f), \sigma(f))$ is the required splitting. (The point here is that once something is given locally, i.e., at a point, we can extend it to a neighborhood if it is "finitely defined", i.e., if we only have to invert finitely many elements of R, or equivalently just one element, to define it.)

$(2) \Rightarrow (3)$: For each p, choose $f = f(p)$ not in p, as in (2), then the $f(p)$ generate the unit ideal; hence a finite number of them suffice, say $(1) = (f_1, \ldots, f_r)$. (This amounts to saying that because $\text{Spec}(R)$ is compact, the open cover $\bigcup_p \text{Spec}(R_{f(p)})$ has a finite subcover $\bigcup_i \text{Spec}(R_{f_i})$.) Now put $S = \prod_i S(f_i)$, $P = \prod_i P(f_i)$, and $\sigma = \prod_i \sigma(f_i)$; then (S, P, σ) is the required splitting. S is faithfully flat by 10.2(b), and S is finitely generated as a module over $\prod_i R_{f_i}$, which in turn is finitely generated as R-algebra.

$(4) \Rightarrow (1)$: Given (4), we have that $A_S = S \otimes_R A$ is a central separable S-algebra, by 2.17, so that A_S is faithfully S-projective and $A_S \otimes_S (A_S)^\circ \to \text{End}_S(A_S)$ is an isomorphism (2.14). Using faithful flatness of S over R, it follows that A is faithfully R-projective, and, from the commutative diagram

$$\begin{array}{ccc} A_S \otimes_S (A_S)^\circ & \longrightarrow & \text{End}_S(A_S) \\ \uparrow\wr & & \uparrow\wr \\ S \otimes_R (A \otimes_R A^\circ) & \longrightarrow & S \otimes_R \text{End}_R(A) \end{array}$$

that $\eta : A \otimes_R A^\circ \to \text{End}_R(A)$ is an isomorphism.

12.4 **Lemma.** *Let* T *be a commutative Noetherian ring and let* ϕ *be an automorphism of the* T-*algebra* $E = \text{End}_T(T^n)$. *There is a monomorphism* $f : T^n \to T^n$ *such that* $(\phi\alpha)f = f\alpha$ *for all* α *in* E. *If* g *is another such, then* $af = bg$ *for non-zero-divisors* a,b *in* T.

Proof. Let K be the localization of R at the multiplicative set S consisting of all non-zero-divisors. Then K is semilocal. (This is a standard fact, whose proof we leave as an exercise: S is the complement of a finite union of prime ideals of R (see ATIYAH & MACDONALD [1], Prop. 4.7) and the localization of R at any such S is semilocal (see BOURBAKI [2], chapter II, §3, No. 5).) Hence if $\bar{\phi}$ denotes the induced automorphism of $\bar{E} = \text{End}_K(K^n)$, there is a K-automorphism $\bar{f}$ of K^n such that $\bar{\phi}(\beta) = \bar{f}\beta\bar{f}^{-1}$ for all β in $\bar{E}$ (exercise; $\text{Pic}(K) = 0$ by 1.13(b) ; then use the remark preceding 7.6). An f as required is then obtained by taking $f = \bar{f}d$ where d = product of denominators in the entries of a matrix representing $\bar{f}$. The determinant of f is a non-zero-divisor, so f is injective by 1.16(c). If $g : T^n \to T^n$ is a monomorphism satisfying $(\phi\alpha)g = g\alpha$ for all α in E, let g' be the classical adjoint of g, so that $gg' = g'g = \det g$. Multiplying $(\phi\alpha)g = g\alpha$ on left and right by g' gives $g'(\phi\alpha) = \alpha g'$, for all α in E, so that $g'f$ is in T, the center of E. Since $gg'f = (\det g)f$ we can take $a = \det g$, $b = g'f$; these are non-zero-divisors because f, g, g' are all injective.

12.5 **Lemma.** *Let* X *and* Y *be submodules of an* R-*module* M, *such that* $X_p = Y_p$ *for every maximal ideal* p *of* R. *Then* $X = Y$.

Proof. Let $Z = X \cap Y$. It suffices to see that the inclusion $Z \hookrightarrow X$ is surjective, for which it's enough to check that the induced inclusion $Z_p \hookrightarrow X_p$ is surjective. But $Z_p = (X \cap Y)_p = X_p \cap Y_p$ (this is easily checked), and the induced map in question is the identity.

12.6 **Lemma.** *Let* T, ϕ, f *be as in* 12.4. *Write* T^N *for the tensor product of* n *copies of* T^n *over* T, *and* $\phi^{(n)}$ (*resp.* $f^{(n)}$) *for the tensor product of* n *copies of* ϕ (*resp.* f) *over* T. *There is an isomorphism* $h : T^N \to T^N$ *such that*
(1) $(\phi^{(n)}\alpha)h = h\alpha$ *for all* α *in* $E = \text{End}_T(T^N)$, *and*
(2) $(\det f)h = f^{(n)}$.

Proof. Note that (2) determines h uniquely since $\det f$ is a non-zero-divisor. Suppose we show $(\det f)T^N = f^{(n)}(T^N)$. Then, given x in T^N, there exists y in T^N with $f^{(n)}(x) = (\det f)y$; y is unique because $\det f$ is a non-zero-divisor; put $h(x) = y$. Then $h : T^N \to T^N$ satisfies (2), and h is obviously surjective; also, $f^{(n)}$ is injective (because f is, and T^n, being free, is flat) so that h is injective too, because $(\det f)h = f^{(n)}$. Finally, from $\phi(\theta)f = f\theta$ for all θ in $\text{End}_T(T^n)$ it follows that $\phi^{(n)}(\alpha)f^{(n)} = f^{(n)}\alpha$ for all α in $E = \text{End}_T(T^N)$; then $(\det f)\phi^{(n)}(\alpha)h = \phi^{(n)}(\alpha)(\det f)h = \phi^{(n)}(\alpha)f^{(n)} = f^{(n)}\alpha = (\det f)h\alpha$ shows that (1) is also satisfied. Thus everything follows if we show $(\det f)T^N = f^{(n)}(T^N)$.

It suffices, by 12.5, to show that $((\det f)T^N)_p = (f^{(n)}(T^N))_p$ at each prime ideal p. Now for each p there is an isomorphism $g(p) : T_p^n \to T_p^n$ such that $\phi_p(\beta) = g(p)\beta g(p)^{-1}$ for all β in $E_p = \text{End}_{T_p}(T_p^n)$, because every automorphism of E_p is inner (T_p is local; see remarks preceding 7.6). Since $(\phi\alpha)f = f\alpha$ for all

α in $E = \text{End}_T(T^n)$, we have $\phi_p(\beta)f_p g(p)^{-1} = f_p \beta g(p)^{-1} = f_p g(p)^{-1}\phi_p(\beta)$ for all β in E_p. Hence $t(p) = f_p g(p)^{-1}$ is in the center T_p of E_p. Therefore $f_p^{(n)} = t(p)^n g(p)^{(n)}$, and $(\det f)_p = \det(f_p) = t(p)^n \det g(p)$. (Here the superscript (n) denotes n-fold tensor product over R_p.) Since $g(p)$ is an automorphism of T_p^n we have $(f^{(n)}(T^N))_p = f_p^{(n)}(T_p^N) = (t(p)^n g(p)^{(n)})T_p^N = t(p)^n T_p^N$. On the other hand, since $\det g(p)$ is invertible we have $((\det f)T^N)_p = (\det(f_p))T_p^N = (t(p)^n \det g(p))T_p^N = t(p)^n(T_p^N)$. Thus $((\det f)T^N)_p = (f^{(n)}(T^N))_p$, and the proof of 12.6 is complete.

12.7 <u>Definition</u>. Let (S, P, σ) be a splitting for an R-algebra A (Definition 12.2). Let χ be the isomorphism of $S \otimes_R S$-algebras defined by the commutative diagram

$$\begin{array}{ccc} S \otimes A \otimes S & \xrightarrow{\sigma \otimes 1} & \text{End}_S(P) \otimes S = \text{End}_{S \otimes S}(P \otimes S) \\ \tau \downarrow & & \downarrow \chi \\ S \otimes S \otimes A & \xrightarrow{1 \otimes \sigma} & S \otimes \text{End}_S(P) = \text{End}_{S \otimes S}(S \otimes P) \end{array}$$

where $\tau(s \otimes a \otimes t) = s \otimes t \otimes a$ and the equalities are the canonical isomorphisms. We will call χ the <u>axe</u> for the splitting (S, P, σ). Note that if P is free, say $P = S^m$, we have $S \otimes P = (S \otimes S)^m = P \otimes S$, so that χ is an automorphism of $\text{End}_T(T^m)$ where $T = S \otimes S$, and the preceding lemmas apply (with $\chi = \phi$).

12.8 <u>Proposition</u>. <u>Let</u> (S, P, Σ) <u>be a splitting for an R-algebra</u> A <u>with</u> S <u>faithfully flat over</u> R <u>(hence</u> A <u>is central separable) and let</u> χ <u>be the axe for the splitting. Assume</u>:

(1) There is an $S \otimes S$-isomorphism $h : \mathcal{P} \otimes S \to S \otimes \mathcal{P}$ such that $(\chi\alpha)h = h\alpha$ for all α in $\mathrm{End}_{S \otimes S}(\mathcal{P} \otimes S)$, and

(2) h is a descent isomorphism, i.e., $h_2 = h_1 h_3$ in the notation of chapter 10.

Then $[\mathcal{A}] \sim 1$ in $B(R)$.

Proof. The results of chapter 10 show that there is an R-module $\mathcal{Q}$ (necessarily faithfully projective) and an S-isomorphism $\eta : S \otimes \mathcal{Q} \to \mathcal{P}$ such that

$$\begin{array}{ccc} S \otimes \mathcal{Q} \otimes S & \xrightarrow{\eta \otimes 1} & \mathcal{P} \otimes S \\ \tau \downarrow & & \downarrow h \\ S \otimes S \otimes \mathcal{Q} & \xrightarrow{1 \otimes \eta} & S \otimes \mathcal{P} \end{array}$$

commutes, where $\tau(s \otimes q \otimes t) = s \otimes t \otimes q$. Define $\rho : S \otimes \mathrm{End}_R(\mathcal{Q}) = \mathrm{End}_S(S \otimes \mathcal{Q}) \to \mathrm{End}_S(\mathcal{P})$ by $\rho(f) = \eta f \eta^{-1}$, then it is easily checked that

$$\begin{array}{ccc} S \otimes \mathrm{End}_R(\mathcal{Q}) \otimes S & \xrightarrow{\rho \otimes 1} & \mathrm{End}_S(\mathcal{P}) \otimes S = \mathrm{End}_{S \otimes S}(\mathcal{P} \otimes S) \\ \tau \downarrow & & \downarrow \chi \\ S \otimes S \otimes \mathrm{End}_R(\mathcal{Q}) & \xrightarrow{1 \otimes \rho} & S \otimes \mathrm{End}_S(\mathcal{P}) = \mathrm{End}_{S \otimes S}(S \otimes \mathcal{P}) \end{array}$$

commutes too, because $\chi(\alpha) = h\alpha h^{-1}$ for all α. It follows from the uniqueness part of 10.7 and 10.9, applied to $(\mathcal{A}, \Sigma)$ and $(\mathrm{End}_R(\mathcal{Q}), \mathcal{P})$, that $\mathcal{A}$ and $\mathrm{End}_R(\mathcal{Q})$ are isomorphic R-algebras; done.

At last:

12.9 Theorem. *For any commutative ring* R, $B(R)$ *is a torsion group*; i.e., *given a central separable* R-*algebra* A, *the* R-*algebras* $A \otimes \ldots \otimes A$ (n *copies*) *and* $\text{End}_R(P)$ *are isomorphic, for some* $n \geq 1$ *and some faithfully projective* R-*module* P.

Proof. We may assume R is Noetherian, by 12.1. Hence we can also assume R is connected, by 5.10 (d) and (i). Then A has constant rank (5.10(e)), say $[A : R] = n^2$ (cf. remark after 3.2). We will show that $[A]^n = 1$ in $B(R)$. The strategy is to apply 12.8 to $\mathcal{A} = A \otimes \ldots \otimes A$ (n factors).

A has a splitting (S, P, σ) with S faithfully flat over R (12.3). Also, we can assume S, $S \otimes S$ and $S \otimes S \otimes S$ are Noetherian, since S can be taken finitely generated as R-algebra. Finally, note that P is free of rank n. (To see this, return to the proof of 12.3: $P(p)$ was free over $S(p)$, say of rank $n(p)$, and because $[A : R] = n^2$ we have $n(p) = n$ for all p.) Now let ϕ be the axe for the splitting. Use 12.4 to find a monomorphism $f : (S \otimes S)^n \to (S \otimes S)^n$ such that $\phi(\alpha)f = f\alpha$ for all α in $\text{End}_{S \otimes S}((S \otimes S)^n)$, and let h be an isomorphism as in 12.6. Thus h is an $S \otimes S$-automorphism of $(S \otimes S)^n \otimes_{S \otimes S} \ldots \otimes_{S \otimes S} (S \otimes S)^n = (S \otimes S)^N$ where $N = n^n$, and $(\phi^{(n)}\alpha)h = h\alpha$ for all α and $(\det f)h = f^{(n)}$.

Now let $\mathcal{A} = A \otimes_R \ldots \otimes_R A$ (n copies). Note that $(S, S^N, \sigma^{(n)})$ splits $\mathcal{A}$, and that the axe for this splitting is $\phi^{(n)}$ (where the superscripts (n) denote n-fold tensor product over S and $S \otimes S$ respectively). According to 12.8 we are done if h, viewed as a map from $S^N \otimes_R S$ to $S \otimes_R S^N$, is a descent isomorphism.

First of all, we have $\phi_2 = \phi_1 \phi_3$. For clearly $\tau_2 = \tau_1 \tau_3$ where $\tau : S \otimes A \otimes S \to S \otimes S \otimes A$ is given by $\tau(s \otimes a \otimes t) = s \otimes t \otimes a$, so that the lefthand triangle in the diagram

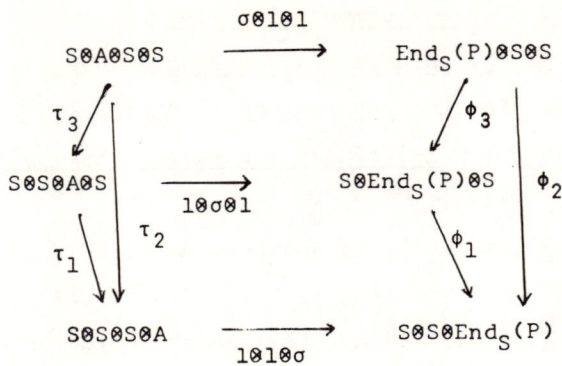

commutes; but ϕ is defined by the commutative diagram

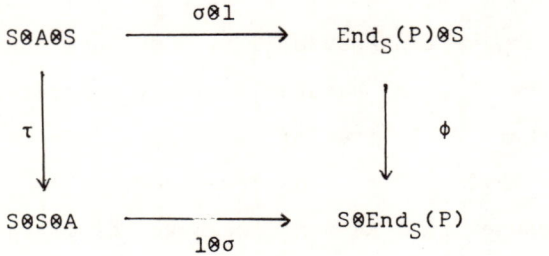

which shows that the three rectangular faces of the prism commute too. Clearly, since $(\phi\alpha)f = f\alpha$ for all α in $\text{End}_{S \otimes S}((S \otimes S)^n)$, we have $(\phi_i \alpha)f_i = f_i \alpha$ for all α in $\text{End}_{S \otimes S \otimes S}((S \otimes S \otimes S)^n)$. Because $\phi_2 = \phi_1 \phi_3$, it follows that $(\phi_2 \alpha)f_1 = \phi_1(\phi_3 \alpha)f_1 = f_1(\phi_3 \alpha)$, and therefore $(\phi_2 \alpha)f_1 f_3 = f_1(\phi_3 \alpha)f_3 = f_1 f_3 \alpha$. On the other hand, $(\phi_2 \alpha)f_2 = f_2 \alpha$, and consequently we can apply the last part of 12.4 (reading $S \otimes S \otimes S$ for T, ϕ_2 for ϕ, f_2 for f, and $f_1 f_3$ for g) to find non-zero-divisors a and b in $S \otimes S \otimes S$ with $af_2 = bf_1 f_3$, and therefore $a^n(\det f_2) = b^n \det(f_1 f_3)$. Now letting (n) denote n-fold tensor product over $S \otimes S \otimes S$, we have also $a^n f_2^{(n)} = b^n f_1^{(n)} f_3^{(n)}$, and multiplying the two equations yields $a^n b^n \det(f_1 f_3) f_2^{(n)} = a^n b^n (\det f_2) f_1^{(n)} f_3^{(n)}$, whence $\det(f_1 f_3) f_2^{(n)} = (\det f_2) f_1^{(n)} f_3^{(n)}$.

Now recall that h was defined by $(\det f)h = f^{(n)}$. It follows that $(\det f_i)h_i = f_i^{(n)}$, and therefore that $(\det f_1)(\det f_2)(\det f_3)h_2 = \det(f_1 f_3)f_2^{(n)} = (\det f_2)f_1^{(n)}f_3^{(n)} = (\det f_1)(\det f_2)(\det f_3)h_1 h_3$. But $\det f_i$ cannot be a zero-divisor in $S \otimes S \otimes S$, because $\det f$ is not a zero-divisor in $S \otimes S$. Hence $h_2 = h_1 h_3$, and we are done.

CHAPTER 13. THE FULL BRAUER GROUP AND CECH COHOMOLOGY.

by Lindsay N. Childs

In chapter 7 it was shown that if S is a Galois extension of R with respect to G, and $Pic(S) = 0$, then $H^2(G,U(S)) \cong B(S/R)$. When S (and hence R) is a field this is very well known. Also well-known for R a field is that every central simple R-algebra is split by a Galois extension S of R. These two results yield a cohomological description, as sketched in chapter 3, of the full Brauer group $B(R)$ of a field R: let $\bar{R}$ be a separable closure of R, and let S_i, i in I, be a family of finite Galois extensions of R contained in $\bar{R}$, with $Gal(S_i/R) = G_i$ and $\varinjlim S_i = \bar{R}$. Then

(13.1) $B(R) = B(\bar{R}/R) \cong \varinjlim H^2(G_i, U(S_i)) = H^2(Gal(\bar{R}/R), U(\bar{R}))$.

When R is no longer a field, this description runs into the obstacle that not every central separable R-algebra need be split by a Galois extension of R. For example, let $R = \mathbb{Z}[\sqrt{2}]$, and let A be a maximal R-order in the quaternion algebra $\Sigma = (\frac{-1,-1}{K})$ over $K = \mathbb{Q}[\sqrt{2}]$. One can show that $[\Sigma]$ becomes trivial in $B(K_p)$ at all finite primes p, but it is non-trivial at each of the two real places. Hence, by 6.35, $[\Sigma]$ comes from a non-trivial element of $B(R)$, and in fact that element is $[A]$ (there is no choice, because $B(R)$ has order 2 by 6.36; cf. the proof of 6.35.) On the other hand, it is folklore that R has no faithful separable extensions except for the trivial ones $R \times \ldots \times R$. (To show this it suffices, by results of JANUSZ [1], to show that every extension L/K of number fields ramifies at some finite prime, when $K = \mathbb{Q}[\sqrt{2}]$. The Hermite-Minkowski theorem gives the corresponding result for K=$\mathbb{Q}$ (see SAMUEL [1], Section 4.3, Theorem 1), and the fact that an analogous proof works for K=$\mathbb{Q}[\sqrt{2}]$ has been in the folklore for some time. Such a proof appears in KNUS & OJANGUREN [5], pp. 91-92.)

An isomorphism of the full Brauer group of R with a cohomology group has to involve taking a limit over a large enough collection of R-algebras to include an algebra to split any given central separable R-algebra. The example above shows that the class of Galois extensions, which suffices when R is a field, is not good enough in general. A suitable collection was constructed in 12.3, but it is convenient to have a larger class:

13.2 <u>Definition</u>. A commutative R-algebra S is <u>étale</u> if

(1) S is faithfully flat as R-module and finitely presented as R-algebra (the latter means there is a surjection $R[X_1,\ldots,X_n] \to S$, for some n , whose kernel is a finitely generated ideal; thus if R is Noetherian it just means finitely generated) and (2) For each prime ideal q of S with, say, $p = q \cap R$, S_q/pS_q is a finite-dimensional separable R_p/pR_p-algebra (hence a finite product of separable finite field extensions of R_p/pR_p).

(This definition is extracted from RAYNAUD [1], Cor. 1, p. 55 ; Prop. 10, p.33 ; and Prop. 11, p. 34 . See also MUMFORD [1], Thm. 3, p. 436.)

The relevant facts about étale algebras which we need are the following:

13.3 __Proposition__. __If__ S __is an étale__ R-__algebra and__ T __is an étale__ S-__algebra then__ T __is an étale__ R-__algebra__.

13.4 __Proposition__. __If__ S __is an étale__ R-__algebra and__ T __is any__ R-__algebra then__ $T \otimes_R S$ __is an étale__ T-__algebra__.

13.5 __Proposition__. __If__ S __and__ T __are étale__ R-__algebras then so is__ $S \otimes_R T$.

13.6 __Proposition__. __The splitting algebras__ S __of__ 12.3(3) __are étale__.

The first three of these can be found in RAYNAUD [1], Prop. 1-3, pp.13-14. The last one, 13.6, is easy if R is Noetherian, and since we will be proving our main result (13.12) for Noetherian R anyway, we will not pursue the general case.

The association of cohomology classes to Brauer classes split by an étale R-algebra S will be given by faithfully flat descent (Chapter 10), so it is natural that the cohomology groups be constructed using the maps defined just below 10.4. We recall the notation used there.

For a commutative R-algebra S , set $S_{(n)} = S \otimes_R \ldots \otimes_R S$ (n copies), $S_{(0)} = R$. Let $d_i : S_{(n)} \to S_{(n+1)}$ (i=1,...,n+1) be defined by $d_i(s_1 \otimes \ldots \otimes s_i \otimes \ldots \otimes s_n) = s_1 \otimes \ldots \otimes s_{i-1} \otimes 1 \otimes s_i \otimes \ldots \otimes s_n$ for $1 \leq i \leq n+1$. Let F be a functor from the category of commutative R-algebras to the category of abelian groups. For an R-algebra T we will use multiplicative notation in F(T) (anticipating such applications as F = U , the group of units). Define $\delta_n : F(S_{(n+1)}) \to F(S_{(n+2)})$ ($n \geq -1$) by

$\delta_n = \prod_{i=1}^{n+2} F(d_i)^{(-1)^i}$. Since $d_{i+1}d_j = d_j d_i$ for all $j \leq i$, it is easily checked that $\delta_{n+1}\delta_n = 0$, so that $\{F(S_{(n+1)}), \delta_n\}_{n \geq -1}$ is a chain complex.

13.7 <u>Definition</u>. Ker δ_n / Im $\delta_{n-1} = H^n(S/R, F)$ is called the n-th <u>Amitsur cohomology group with coefficients in</u> F. Ker $\delta_n = Z^n$ is the group of n-cocycles; Im $\delta_{n-1} = B^n$ is the group of n-coboundaries.

In low dimensions, $H^n(S/R, F)$ looks as follows:
$H^0(S/R,F) = Z^0/B^0$ where $Z^0 = \{\alpha \text{ in } F(S) | F(d_1)(\alpha) = F(d_2)(\alpha)\}$ and $B^0 = \{\alpha \text{ in } F(S) | \alpha = F(d_1)(\beta), \beta \text{ in } F(R)\}$. Since $F(d_1)$ is the trivial map on $F(R)$, B^0 is trivial.
$H^1(S/R,F) = Z^1/B^1$ where $Z^1 = \{\alpha \text{ in } F(S \otimes S) | F(d_1)(\alpha)F(d_3)(\alpha) = F(d_2)(\alpha)\}$ and $B^1 = \{\alpha \text{ in } F(S \otimes S) | \alpha = F(d_1)(\beta)F(d_2)(\beta)^{-1}, \beta \text{ in } F(S)\}$.

We will need non-abelian versions of these groups. The definitions which follow are adapted from HOOBLER [4], chapter 1.

13.8 <u>Definition</u>. Let F be a functor from commutative R-algebras to (not necessarily abelian) groups. Define the cohomology group $H^0(S/R,F)$ to be $\{\alpha \text{ in } F(S) | F(d_1)(\alpha) = F(d_2)(\alpha)\}$. Let $Z^1(S/R,F) = \{\alpha \text{ in } F(S \otimes S) | F(d_1)(\alpha)F(d_3)(\alpha) = F(d_2)(\alpha)\}$ and define an equivalence relation on $Z^1(S/R, F)$ by: $\alpha \sim \alpha'$ iff there exists β in $F(S)$ with $\alpha = F(d_1)(\beta)\alpha F(d_2)(\beta)^{-1}$. The pointed set $Z^1(S/R, F)/\sim$ of equivalence classes is denoted $H^1(S/R, F)$.

Of course, 13.7 and 13.8 agree when they both apply. It is convenient to introduce an abbreviated notation: if α is in

$F(S_{(n)})$ we set $\alpha_i = F(d_i)(\alpha)$ in $F(S_{(n+1)})$. Thus we will write $\alpha \sim \alpha'$ iff $\alpha' = \beta_1 \alpha \beta_2^{-1}$. Note that this is consistent with the notation of chapters 10 and 12.

That Amitsur cohomology is appropriate for describing the Brauer group is suggested by the following result, which relates the results of Chapter 7 to what follows.

13.9 <u>Proposition</u>. <u>Let</u> S <u>be a Galois extension of</u> R <u>with group</u> G. <u>Let</u> F <u>be a functor from commutative R-algebras to abelian groups such that</u> $F(S_1 \times S_2) = F(S_1) \oplus F(S_2)$ <u>for any R-algebras</u> S_1, S_2. <u>Let</u> $C^n(G,X)$ <u>denote the set of set functions from</u> $G \times \ldots \times G$ (n times) <u>to</u> X. <u>Then the maps</u> $\theta_n : S_{(n+1)} \to C^n(G,S)$ <u>given by</u> $\theta_n(s_1 \otimes \ldots \otimes s_{n+1})(\sigma_1, \ldots, \sigma_n) = s_1 \sigma_1(s_2) \sigma_1 \sigma_2(s_3) \ldots (\sigma_1 \ldots \sigma_n)(s_{n+1})$ <u>induce coboundary-preserving isomorphisms</u> $F(S_{(n+1)}) \to F(C^n(G,S)) \simeq C^n(G,F(S))$ <u>and therefore isomorphisms</u> $H^n(S/R,F) \simeq H^n(G,F(S))$ <u>for all</u> $n \geq 0$.

For details see CHASE, HARRISON & ROSENBERG [1], Theorem 5.4. The same result holds for $n = 0,1$ if F maps to (not necessarily abelian) groups and non-abelian group cohomology is defined as in SERRE [2], pp. 131-134.

Amitsur cohomology behaves well with respect to natural transformations of functors, and very well with respect to R-algebra maps. We prove the latter for $n = 0,1$ and F non-abelian, leaving the abelian case where $n \geq 2$ as an exercise. Compare CHASE & ROSENBERG [1], beginning of §3.

13.10 <u>Theorem</u>. (HOOBLER [4], Prop. 1.2) <u>Let</u> F <u>be a functor from</u> R-<u>algebras to groups and let</u> $f,g : S \to T$ <u>be</u> R-<u>algebra homomorphisms. Then the induced maps</u> f_*, g_*:

$H^n(S/R,F) \to H^n(T/R,F)$ (n=0,1) are the same.

Proof. Since F is a functor, f and g induce maps $F(f\otimes\ldots\otimes f):F(S_{(n)}) \to F(T_{(n)})$, etc., and hence maps f_* and g_* from $H^n(S/R,F)$ to $H^n(T/R,F)$. We show that $f_* = g_*$ separately for $n = 0,1$.

For $n = 0$, define $h: S\otimes S \to T$ by $h(s_1\otimes s_2) = f(s_1)g(s_2)$. Then $hd_1 = g$, $hd_2 = f$. If x is in $H^0(S/R,F) \subseteq F(S)$ then $F(d_1)(x) = F(d_2)(x)$, so $F(g)x = F(h)F(d_1)(x) = F(h)F(d_2)(x) = F(f)x$.

For $n = 1$, define $h^i: S_{(3)} \to T_{(2)}$ by $h^1(s_1\otimes s_2\otimes s_3) = f(s_1)g(s_2)\otimes g(s_3)$, $h^2(s_1\otimes s_2\otimes s_3) = f(s_1)\otimes f(s_2)g(s_3)$. Then, as is easily checked, we have $h^1 d_1 = g\otimes g$, $h^1 d_2 = f\otimes g$, $h^1 d_3 = d_2 h$, $h^2 d_1 = d_1 h$, $h^2 d_2 = f\otimes g$ and $h^2 d_3 = f\otimes f$. Now if x is in $Z^1(S/R,F)$, x is in $F(S\otimes S)$ and $F(d_2)(x) = F(d_1)(x)F(d_3)(x)$. Then on the one hand, $F(f\otimes g)(x) = F(h^1 d_2)(x) = F(h^1 d_1)(x)F(h^1 d_3)(x) = F(g\otimes g)(x)F(d_2 h)(x)$; on the other hand, $F(f\otimes g)(x) = F(h^2 d_2)(x) = F(h^2 d_1)(x)F(h^2 d_3)(x) = F(d_1 h)(x)F(f\otimes f)(x)$. Setting $y = F(h)(x)$ we get $F(g\otimes g)(x) = F(d_1)(y)\cdot F(f\otimes f)(x)\cdot F(d_2)(y)^{-1}$. Thus the maps induced by f and g on $H^n(S/R,F)$ are the same.

This last result implies that if we let $E(R)$ be the set of isomorphism classes $[S]$ of étale R-algebras S (this will be a set since the cardinality of such an algebra S is either countable or the same as that of R, because S is finitely generated as R-algebra), and put a partial order on $E(R)$ by $[S] \leq [T]$ iff the set of R-algebra maps from S to T is non-empty, then $E(R)$ is a directed (13.5) partially ordered set on which, when viewed as a category, $H^n(\cdot/R,F)$ is

a functor. Thus $\varinjlim H^n(S/R,F)$, where the limit is taken over
[S] in E(R) , is defined.

13.11 <u>Definition</u>. $H^n_{et}(R,F) = \varinjlim_{[S] \text{ in } E(R)} H^n(S/R,F)$ is
called the n-th Cech cohomology group with coefficients in F ,
with respect to the étale topology.

A result of GROTHENDIECK [1] suggests the following as a
suitable generalization of 13.1 :

13.12 <u>Theorem</u>. <u>There is an injective homomorphism</u> λ <u>from</u>
$B(R)$ <u>to</u> $H^2_{et}(R,U)$.

We will describe a proof of this theorem, based on the
results of chapters 10 and 12, assuming that R is Noetherian
and connected. We leave to the reader the exercise of seeing
whether 12.1 allows the proof to go through for general R .

The strategy of the proof is as follows. For each n
we have the short exact sequence $1 \to U \to GL_n \to PGL_n \to 1$ (this
can be taken as the definition of PGL_n) . We view $PGL_n(R)$
as the group of inner automorphisms of the matrix ring $(R)_n$.
Applying Amitsur cohomology to the above sequence, we obtain
maps $H^1(S/R,GL_n) \xrightarrow{\bar{j}} H^1(S/R,PGL_n) \xrightarrow{\partial} H^2(S/R,U)$ as follows:
$\bar{j}$ is induced from $j: GL_n(S \otimes S) \to PGL_n(S \otimes S)$ in the obvious way;
∂ is defined by taking ϕ in $PGL_n(S \otimes S)$ representing a class
in $H^1(S/R,PGL_n)$, viewing ϕ as conjugation by some f in
$GL_n(S \otimes S)$, and observing that there exists u in $U(S \otimes S \otimes S)$ so
that $ud_2(f) = d_1(f)d_3(f)$; then ∂ of the class of ϕ is
the class of u . The sequence $\ldots H^1(S/R,GL_n) \xrightarrow{\bar{j}} H^1(S/R,PGL_n) \xrightarrow{\partial}$
$H^2(S/R,U)$ of pointed sets is in fact exact: exactness at

$H^1(S/R, PGL_n)$ is essentially 13.16 and 13.18 below. Now $\bar{\lambda}$ is given as follows: use the faithfully flat descent of chapter 10 to associate to a central separable R-algebra A of rank n^2 split by S an element of $H^1(S/R, PGL_n)$, map that element by ∂ over to $H^2(S/R, U)$, then into the limit $H^2_{et}(R, U)$.

We begin the details of constructing $\bar{\lambda}$.

Let A be a central separable R-algebra of rank n^2. By 12.3 and 13.6 we can find an étale R-algebra S which splits A into an n×n matrix ring. Thus to A corresponds $(End_S(S^n), \phi)$ where ϕ, the axe for the splitting, is an S⊗S-algebra automorphism of $End_{S \otimes S}((S \otimes S)^n)$. This element ϕ satisfies $\phi_2 = \phi_1 \phi_3$ (see proof of 12.9), hence is a one-cocycle representing a class in $H^1(S/R, Aut(End_{(\ )-alg}((\ )^n)))$. If ϕ were an inner automorphism of $End_{S \otimes S}((S \otimes S)^n)$ then ϕ would represent a class in $H^1(S/R, PGL_n)$, and according to our outline of the construction of $\bar{\lambda}$ we would be in good shape. Unfortunately, it need not be inner, and, just as in chapter 7, rank one projectives interfere.

There is, by Morita theory, an exact sequence of functors on the category of R-algebras

$$1 \to PGL_n \to Aut(End_{(\ )-alg}((\ )^n)) \xrightarrow{J} tPic$$

where tPic(T) is the torsion subgroup of Pic(T). (See ROSENBERG & ZELINSKY [2], or BASS [1], Prop. 7.3 of chapter II.) In order to define $\bar{\lambda}$ one must circumvent the obstruction, in tPic, to axes being inner.

One knows that it is possible to do this. Let S be an étale R-algebra which splits A, with axe ϕ in $\mathrm{Aut}(\mathrm{End}_{S \otimes S}((S \otimes S)^n))$. Let J_ϕ be the rank one projective $S \otimes S$-module corresponding to ϕ, i.e., the obstruction to its innerness. Then there is an étale $S \otimes S$-algebra W such that $J_\phi \otimes_{S \otimes S} W$ is free; one such can be constructed as follows. For each prime ideal p of $S \otimes S$, $J_\phi \otimes_{S \otimes S} (S \otimes S)_p$ is free, so just as in 12.3 one can find $f(p)$ in $S \otimes S$ but not in p such that $J_\phi \otimes_{S \otimes S} (S \otimes S)_{f(p)}$ is free. Choose such $f(p)$ for each p; then the ideal generated by all the $f(p)$ is the unit ideal, so that $1 = \Sigma a_i f_i$ for certain $f_i = f(p_i)$ ($i=1,\ldots,n; a_i$ in $S \otimes S$). Then $J_\phi \otimes_{S \otimes S} (S \otimes S)_{a_i f_i} = J_\phi \otimes_{S \otimes S} (S \otimes S)_{f_i} \otimes (S \otimes S)_{f_i} (S \otimes S)_{f_i a_i}$ is free, and so $J_\phi \otimes_{S \otimes S} W$ is free where $W = \Sigma (S \otimes S)_{f_i a_i}$. W is an étale $S \otimes S$-algebra by 13.6.

Now we invoke a difficult theorem of M. Artin (Theorem 4.1 of ARTIN [2]) which says that given an étale $S_{(n)}$-algebra W, there is an etale S-algebra T and a map $W \to T_{(n)}$ making

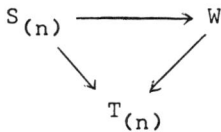

commute, where $S_{(n)} \to T_{(n)}$ is the canonical map. T also has the following properties:

13.13 <u>Proposition.</u> <u>With notation as above, we have</u>:
(a) T <u>is étale over</u> R (by 13.3);
(b) $T \otimes_R A \simeq \mathrm{End}_T(T^n)$, <u>the map being induced from the</u>

<u>splitting by</u> S <u>of</u> A ; and

(c) <u>The axe</u> ϕ <u>for the splitting of</u> A <u>by</u> T (part (b)) <u>is an inner automorphism of</u> $\text{End}_{T \otimes T}((T \otimes T)^n)$.

The operative part of 13.13 is the word "inner" in part (c) . Thus, using Artin's Theorem, given a central separable R-algebra A there is an étale splitting algebra T for A <u>for which the axe is inner</u>. Hence we can, and therefore will, assume that any splitting algebra satisfies the properties of 13.13.

It would be a worthwhile problem to see if one can define directly (i.e. without recourse to Artin's theorem) an étale splitting algebra for a central separable R-algebra A such that the axe for the splitting is inner. This would simplify the proof of torsionness in chapter 12, e.g. by avoiding part of 12.4. We'll return to this point later (cf. 13.20(a)).

We can now define $\bar{\lambda}$. Given a central separable R-algebra A of rank n^2 we choose T satisfying 13.13 and get an axe

$$\phi: \text{End}_{T \otimes T}((T \otimes T)^n) \to \text{End}_{T \otimes T}((T \otimes T)^n)$$ which is inner:

$\phi(\alpha) = f \alpha f^{-1}$ for some invertible f in $\text{End}_{T \otimes T}((T \otimes T)^n)$. Thus [ϕ] is in $H^1(T/R, \text{PGL}_n)$, and since $\phi_2 = \phi_1 \phi_3$, there is a unit u of $T \otimes T \otimes T$ such that $u f_2 = f_1 f_3$. (This can be seen by applying the last part of 12.4 in the same way as in the proof of 12.9 : using the fact that f, f_1, f_2, f_3 are all invertible it follows that $a = \det(f_1 f_3)$ is invertible, hence b is invertible and we can set $u = a/b$ in the formula $a f_2 = b f_1 f_3$ arising in the latter part of the proof of 12.9.)

The element u in $U(T \otimes T \otimes T)$ is easily seen to be a two-cocycle, hence yields an element of $H^2(T/R, U)$.

13.14 <u>Definition</u>. The map $\bar{\lambda} : B(R) \to H^2_{et}(R,U)$ is defined by sending the class of A to the image in $H^2_{et}(R,U)$ of the class of u in $H^2(T/R,U)$.

The rest of this chapter is devoted to verifying that $\bar{\lambda}$ is a well-defined injective homomorphism. To do this, let λ denote the map described from the set of isomorphism classes of central separable R-algebras to $H^2_{et}(R,U)$. We show:

13.15 <u>Proposition</u>. λ <u>is well-defined on each isomorphism class.</u>

13.16 <u>Proposition</u>. λ <u>is trivial on central separable algebras which are trivial in</u> $B(R)$.

13.17 <u>Proposition</u>. $\lambda(A \otimes_R B) = \lambda(A) \cdot \lambda(B)$.

13.18 <u>Proposition</u>. <u>If</u> $\lambda(A)$ <u>is trivial then</u> $[A]$ <u>is trivial in</u> $B(R)$.

<u>Proof of 13.15</u>. We show first that

(a) If A is split by S in such a way that the axe ϕ is inner, then the class of ϕ in $H^1(S/R, PGL_n)$ depends only on the isomorphism class of A . Having done that it is a standard procedure to check :

(b) The map $H^1(S/R, PGL_n) \xrightarrow{\partial} H^2(S/R, U)$, described just below 13.12, is well-defined.

Finally we show

(c) If A is split by S and by T, yielding u_S, u_T in $H^2(S/R,U)$, $H^2(T/R,U)$ respectively, then $\varinjlim u_S = \varinjlim u_T$ in $H^2_{et}(R,U)$.

Proof of (a). Recall that ϕ was defined by the commutative diagram

$$\begin{array}{ccc} S\otimes A\otimes S & \xrightarrow{\sigma\otimes 1} & \text{End}_S(S^n)\otimes S \simeq \text{End}_{S\otimes S}((S\otimes S)^n) \\ \tau\downarrow & & \downarrow\phi \\ S\otimes S\otimes A & \xrightarrow{1\otimes \sigma} & S\otimes \text{End}_S(S^n) \simeq \text{End}_{S\otimes S}((S\otimes S)^n) \end{array}$$

where $\tau(s\otimes a\otimes t) = s\otimes t\otimes a$. We show first that the class of ϕ in $H^1(S/R,PGL_n)$ doesn't depend on the particular splitting isomorphism σ for A.

Let σ, σ' be splittings for A whose axes ϕ, ϕ' are both inner automorphisms of $\text{End}_{S\otimes S}((S\otimes S)^n)$. Put $\psi = \sigma'\sigma^{-1}$ in $\text{Aut}(\text{End}_S(S^n))$. By passing to a further extension if necessary, we can assume ψ is inner, hence in $PGL_n(S)$. Then $\psi_1\phi\psi_2^{-1} = \sigma_1'\sigma_1^{-1}\phi\sigma_2\sigma_2'^{-1} = \sigma_1'\tau\sigma_2'^{-1} = \phi'$, which shows that the class of ϕ is independent of σ.

Now if $\theta: A' \to A$ is an R-algebra isomorphism then $\sigma' = \sigma\theta$ splits A' when σ splits A. The axe for this splitting of A' is clearly the same as the axe for A corresponding to σ. Since the class of the axe is independent of the particular splitting chosen, this proves (a).

Part (b) of 13.15 we leave as an exercise.

Proof of (c): If A is split by S and T so that the

axes ϕ_S, ϕ_T are both inner, then A is also split by $S \otimes T$ and we have two induced axes, both inner automorphisms of $\text{End}_{S \otimes T \otimes S \otimes T}((S \otimes T \otimes S \otimes T)^n)$. By part(a), these give the same element of $H^1(S \otimes T/R, \text{PGL}_n)$. So it suffices to show that the map $H^1(S/R, \text{PGL}_n) \xrightarrow{\partial} H^2(S/R, U)$ is functorial in S, an exercise we leave to the reader.

Proof of 13.16. We suppose $A = \text{End}_R(Q)$ for some faithfully projective R-module Q of rank n. We show that λ is trivial on A by showing that the axe ϕ for A in $H^1(S/R, \text{PGL}_n)$ comes from $H^1(S/R, \text{GL}_n)$.

Let S be an étale R-algebra such that there is an isomorphism $\rho: S \otimes Q \to S^n$ of S-modules. Define an automorphism g of $(S \otimes S)^n$ via commutativity of the diagram

(13.19)
$$\begin{array}{ccc} S \otimes Q \otimes S & \xrightarrow{\rho \otimes 1} & S^n \otimes S \simeq (S \otimes S)^n \\ \tau' \downarrow & & \downarrow g \\ S \otimes S \otimes Q & \xrightarrow{1 \otimes \rho} & S \otimes S^n \simeq (S \otimes S)^n \end{array}$$

where $\tau'(s \otimes x \otimes t) = s \otimes t \otimes x$ (cf. 12.7) and the isomorphisms $S^n \otimes S \simeq (S \otimes S)^n \simeq S \otimes S^n$ are the canonical ones. Then, by the exercise just before 10.7, $g_2 = g_1 g_3$. Thus g represents a class in $H^1(S/R, \text{GL}_n)$. Now define a splitting σ of A by S as follows: compose the natural map $S \otimes \text{End}_R(Q) \simeq \text{End}_S(S \otimes Q)$ with the map $\text{End}_S(S \otimes Q) \to \text{End}_S(S^n)$, call it $C(\rho)$, given by inner automorphism by ρ: $C(\rho)\alpha = \rho \alpha \rho^{-1}$. Then we can describe the axe for this splitting σ: apply $\text{End}_{S \otimes S}(-)$ to 13.19 to get

$$
\begin{CD}
S \otimes \operatorname{End}_R(Q) \otimes S \cong \operatorname{End}_{S \otimes S}(S \otimes Q \otimes S) @>{C(\rho \otimes 1)}>> \operatorname{End}_{S \otimes S}(S^n \otimes S) \cong \operatorname{End}_{S \otimes S}((S \otimes S)^n) \\
@V{\tau}VV @VV{C(\tau')}V @VV{C(g)}V \\
S \otimes S \otimes \operatorname{End}_R(Q) \cong \operatorname{End}_{S \otimes S}(S \otimes S \otimes Q) @>>{C(1 \otimes \rho)}> \operatorname{End}_{S \otimes S}(S \otimes S^n) \cong \operatorname{End}_{S \otimes S}((S \otimes S)^n)
\end{CD}
$$

The righthand square commutes because 13.19 does, and obviously the lefthand square commutes too; thus the axe for the splitting of $\operatorname{End}_R(Q)$ is $C(g)$. Now the map from $H^1(S/R, PGL_n)$ to $H^2(S/R, U)$ associates to an axe ϕ the class of an element u such that, if f in $GL_n(S \otimes S)$ satisfies $\phi(\alpha) = f \alpha f^{-1}$ (i.e., $\phi = C(f)$), then $u f_2 = f_1 f_3$. We choose here $f = g$; then $u = 1$, as claimed.

Proof of 13.17. Let A and B be central separable R-algebras, and suppose that S splits them both. Let ϕ_A in $H^1(S/R, PGL_n)$ and ϕ_B in $H^1(S/R, PGL_m)$ be the corresponding axes. Then S splits $A \otimes B$, and the corresponding axe is $\phi_A \otimes \phi_B$, as one sees by constructing a very large commutative diagram. Now ϕ_A and ϕ_B are inner automorphisms, say by f^A and f^B respectively, and hence $\phi_A \otimes \phi_B$ is inner automorphism, by the automorphism $f^A \otimes f^B$ of $(S \otimes S)^n \otimes_{S \otimes S} (S \otimes S)^m$. Then $(f_1^A \otimes f_1^B)(f_3^A \otimes f_3^B) = (u_A \otimes u_B)(f_2^A \otimes f_2^B)$, where $u_A \otimes u_B$ is in the center $U(S_{(3)})$ of $\operatorname{End}_{S_{(3)}}((S_{(3)})^n \otimes_{S_{(3)}} (S_{(3)})^m)$ (recall that $S_{(3)} = S \otimes S \otimes S$.) So $u_A \otimes u_B = u_A u_B$. Thus the element of $H^2(S/R, U)$ corresponding to $A \otimes B$ is $[u_A u_B] = [u_A] \cdot [u_B]$, the product of the elements corresponding to A and B. This proves 13.17. We now know that $\bar{\lambda} : B(R) \to H^2_{et}(R, U)$ is a well-defined homomorphism; 13.18 says it is injective.

Proof of 13.18. If $\lambda(A)$ is trivial in $H^2_{et}(R,U)$, there is an étale R-algebra S which splits A such that $\lambda(A)$ is trivial in $H^2(S/R,U)$. This means that if ϕ is the axe for a splitting $\sigma: S \otimes A \to End_S(S^n)$, with $\phi(\alpha) = f\alpha f^{-1}$ and $uf_2 = f_1 f_3$, then u is a coboundary, i.e., $u = \delta(v) = v_2^{-1} v_1 v_3$ for some v in $U(S \otimes S)$. But then, putting $h = fv^{-1}$, we have $\phi(\alpha) = h\alpha h^{-1}$, since f in $GL_n(S \otimes S)$ may be altered by an element of $U(S \otimes S)$. A straightforward calculation then shows that $h_2 = h_1 h_3$, and by 12.8 [A] is therefore trivial in B(R). This proves 13.18, and thereby also 13.12.

13.20 Remarks. (a) The fact that the proof of 13.12 could be borrowed so efficiently from chapter 12 illustrates the extent to which the Knus-Ojanguren proof of torsionness of B(R) is really a cohomological argument. In fact, what the proof of 12.9 shows is that if A is a central separable R-algebra of rank n^2, split by S, with axe ϕ, with $\phi(\alpha) = f\alpha f^{-1}$ and $uf_2 = f_1 f_3$, so that $\bar\lambda[A] = [u]$, then $[u]^n = [u^n] = \delta(\det f)$. A comparison of chapters 12 and 13 exhibits two subtle aspects of the Knus-Ojanguren proof; one is that by using 12.6 one is able to circumvent Artin's theorem, by not requiring _inner_ axes. It sufficed instead to know (by 12.4) that there is a monomorphism $f:(S \otimes S)^n \to (S \otimes S)^n$ with $\phi(\alpha)f = f\alpha$ for α in $End_{S \otimes S}((S \otimes S)^n)$, and then to show that $(\det f)h = f^{(n)}$ implies $\phi^{(n)}(\alpha)h = h\alpha$ where in fact h is invertible ($\det h = 1$) and is a cocycle in $GL_n(S \otimes S)$. That h turns out to be a cocycle is the other subtle aspect!

(b) It is natural to ask: when is $\bar\lambda: B(R) \hookrightarrow H^2_{et}(R,U)$ also onto? The answer is: not always. A corollary to the theorem of Artin cited above (ARTIN [2], Cor. 4.2) implies that $H_{et}^2(R,U)$ is isomorphic to the étale sheaf cohomology used in the analysis of the Brauer group in GROTHENDIECK [1], [2], [3]; thus Grothendieck's observations (in [2], section 2) on this question apply. In particular, he points out that $H_{et}^2(R,U)$ need not be torsion. Whether $B(R)$ is isomorphic to the torsion subgroup of $H_{et}^2(R,U)$ is apparently unknown. The reader who has reached this far in these notes should start confronting Grothendieck's three lectures, so we omit further discussion.

BIBLIOGRAPHY

The bibliography that follows includes those items we have referred to in the text of these notes, and in addition contains an expansive list of related material. The main criterion for inclusion has been that the work deal with either the Brauer group or with central separable algebras. The task of including even the recent papers on such related topics as separable algebras or Galois extensions would have been impractically monumental, and an attempt to deal with generalizations of central separable algebras (e.g. to non-commutative base rings) would have quickly led too far afield. As a compromise, a few references to work not specifically dealing with the Brauer group or with central separable algebras have been left in, to serve as a starting point for someone who would care to find out what work has been done in this or that direction. The bibliography of DEMEYER & INGRAHAM [1] may be considered a complement to ours in this respect. We have also included references to some topics which are not touched upon in the text itself, notably work on the Schur subgroup of the Brauer group, since we felt this might be useful to anyone interested in seeing some applications of the Brauer group. Although some of our references trace back to work which appeared barely too late to be included in the 1936 comprehensive bibliography of DEURING [2], the completeness of our compilation generally decreases as one goes back through the years, a situation made necessary by our desire to keep the task of compiling this list within bounds.

ALBERT. A.A.
- [1] Normal division algebras of degree p^e over F of characteristic p. Trans. Amer. Math. Soc. 39 (1936), 183-188.
- [2] Simple algebras of degree p^e over a centrum of characteristic p, Trans. Amer. Math. Soc. 40 (1936), 112-126.
- [3] Non-cyclic algebras with pure maximal subfields, Bull. Amer. Math. Soc. 44 (1938), 576-579.
- [4] A note on normal division algebras of prime degree, Bull. Amer. Math. Soc. 44 (1938), 649-652.
- [5] Structure of Algebras, Amer. Math. Soc. Colloquium Publications, Vol. 24, 1939.
- [6] On p-adic fields and rational division algebras, Ann. of Math. 41 (1940), 674-693.
- [7] Division algebras over a function field, Duke Math. J. 8 (1941), 750-762.
- [8] On associative division algebras of prime degree, Proc. Amer. Math. Soc. 16 (1965), 799-802.
- [9] New results on associative division algebras, J. Algebra 5 (1967), 110-132.
- [10] Tensor products of quaternion algebras, Proc. Amer. Math. Soc. 35 (1972), 65-66.

AMITSUR, S.A.
- [1] La représentation d'algèbres centrales simples, C.R. Acad. Sci. Paris Sér. A, 230 (1950), 902-904.
- [2] Construction d'algèbres contrales simples, C.R. Acad. Sci. Paris Sér. A, 230 (1950), 1026-1028.
- [3] Differential polynomial and division algebras, Ann. of Math. 59 (1954), 245-278.
- [4] Generic splitting fields of central simple algebras, Ann. of Math. 62 (1955), 8-43.
- [5] Some results on central simple algebras, Ann. of Math. 63 (1956), 285-293.
- [6] Simple algebras and cohomology groups of arbitrary fields, Trans. Amer. Math. Soc. 90 (1959), 73-112.
- [7] Homology groups and double complexes for arbitrary fields, J. Math. Soc. Japan 14 (1962), 1-25.
- [8] On central division algebras, Israel J. Math. 12 (1972), 408-420.
- [9] Polynomial identities and Azumaya algebras, J. Algebra 27 (1973), 117-125.

ARTIN, E., NESBITT, C.J. and THRALL, R.M.
- [1] Rings with Minimum Condition, Univ. of Michigan Press, Ann Arbor, 1944.

ARTIN, M.
- [1] Azumaya algebras and finite dimensional representations of rings, J. Algebra 11 (1969), 532-563.
- [2] On the joins of hensel rings, Advances in Math. 7 (1971), 282-296.

ARTIN, M. and MUMFORD, D.
[1] Some elementary examples of unirational varieties which are not rational, Proc. London Math. Soc. (3rd Series) 25 (1972), 75-95.

ATIYAH, M.F., BOTT, R. and SHAPIRO, A.
[1] Clifford Modules, Topology 3 Suppl. 1 (1964), 3-38.

ATIYAH, M.F. and MACDONALD, I.G.
[1] Introduction to Commutative Algebra, Addison-Wesley, Reading, Mass., 1969.

AUSLANDER, B.
[1] The Brauer group of a ringed space, J. Algebra 4 (1966), 220-273.
[2] Central separable algebras which are locally endomorphism rings of free modules, Proc. Amer. Math. Soc. 30 (1971), 395-404.

AUSLANDER, M. and BRUMER, A.
[1] Brauer groups of discrete valuation rings, Nederl. Akad. Wetensch. Proc. Ser. A, 71 (1968), 286-296.

AUSLANDER, M. and GOLDMAN, O.
[1] Maximal orders, Trans. Amer. Math. Soc. 97 (1960), 1-24.
[2] The Brauer group of a commutative ring, Trans. Amer. Math. Soc. 97 (1960), 367-409.

AX, J.
[1] A field of cohomological dimension 1 which is not C_1, Bull. Amer. Math. Soc. 71 (1965), 717.
[2] Proof of some conjectures on cohomological dimension, Proc. Amer. Math. Soc. 16 (1965), 1214-1221.

AZUMAYA, G.
[1] Galois theory for uni-serial rings, J. Math. Soc. Japan 1 (1949), 130-146.
[2] On maximally central algebras, Nagoya Math. J. 2 (1951), 119-150.
[3] The Brauer group of an adèle ring, NSF Advanced Sci. Sem., Bowdoin College, 1966.

AZUMAYA, G. and NAKAYAMA, T.
[1] On absolutely uni-serial algebras, Jap. J. Math. 19 (1948), Transaction 26.

BARR, M. and KNUS, M.-A.
[1] Extensions of derivations, Proc. Amer. Math. Soc. 28 (1971), 313-314.

BASS, H.
- [1] Topics in Algebraic K-Theory, Lectures in Mathematics and Physics No. 41, Tata Institute of Fundamental Research, Bombay, 1967.
- [2] Algebraic K-Theory, Benjamin, New York, 1968.
- [3] Libération des modules projectifs sur certaines anneaux de polynômes, Sém. Bourbaki, June 1974.

BENARD, M.
- [1] Quaternion constituents of group algebras, Proc. Amer. Math. Soc. 30 (1971), 217-219.
- [2] The Schur subgroup I, J. Algebra 22 (1972), 374-377.

BENARD, M. and SCHACHER, M.M.
- [1] The Schur subgroup II, J. Algebra 22 (1972), 378-385.

BERKSON, A.J. and McCONNELL, A.
- [1] On inflation-restriction exact sequences in group and Amitsur cohomology, Trans. Amer. Math. Soc. 141 (1969), 403-413.

BOURBAKI, N.
- [1] Algèbre, Vol. II, Chapter 8, Hermann, Paris, 1958.
- [2] Algèbre Commutative, Chapters 1-7, Hermann, Paris, 1965.

BRAUER, R.
- [1] On sets of matrices with coefficients in a division ring, Trans. Amer. Math. Soc. 49 (1941), 502-548.

BRUMER, A. and ROSEN, M.
- [1] On the size of the Brauer group, Proc. Amer. Math. Soc. 19 (1968), 707-711.

BUMBY, R.T. and DOBBS, D.E.
- [1] Amitsur cohomology of quadratic extensions: formulas and number-theoretic examples, To appear.

CARTAN, H. and EILENBERG, S.
- [1] Homological algebra, Princeton University Press, Princeton, 1956.

CASSELS, J. and FROHLICH, A.
- [1] Algebraic Number Theory, Thompson Book Company, Washington, D.C., 1967.

CHASE, S.U.
- [1] Some remarks on forms of algebras, To appear.

CHASE, S.U., HARRISON, D.K. and ROSENBERG, A.
- [1] Galois theory and Galois cohomology of commutative rings, Mem. Amer. Math. Soc. 52 (1968), 1-19.

CHASE, S.U. and ROSENBERG, A.
- [1] Amitsur cohomology and the Brauer group, Mem. Amer. Math. Soc. 52 (1968), 20-65.
- [2] Centralizers and forms of algebras, To appear.

CHILDS, L.N.
- [1] On projective modules and automorphisms of central separable algebras, Canad. J. Math. 21 (1969), 44-53.
- [2] The exact sequence of low degree and normal algebras, Bull. Amer. Math. Soc. 76 (1970), 1121-1124.
- [3] On normal Azumaya algebras and the Teichmuller cocycle map, J. Algebra 23 (1972), 1-17.
- [4] Brauer groups of affine rings, Ring Theory - Proceedings of the Oklahoma Conference, Lecture Notes in Pure and Applied Mathematics, Marcel Dekker Inc., New York, 1974, 83-92.
- [5] Mayer-Vietoris sequences and Brauer groups of non-normal domains, To appear.
- [6] The Brauer group of graded Azumaya algebras II: graded Galois extensions, To appear.

CHILDS, L.N. and DEMEYER, F.R.
- [1] On automorphisms of separable algebras, Pacific J. Math. 23 (1967), 25-34.

CHILDS, L.N., GARFINKEL, G. and ORZECH, M.
- [1] The Brauer group of graded Azumaya algebras, Trans. Amer. Math. Soc. 175 (1973), 299-326.
- [2] On the Brauer group and factoriality of normal domains, To appear.

COURTER, R.C.
- [1] The dimension of maximal commutative subalgebras of K_n, Duke Math. J. 32 (1965), 225-232.

CURTIS, C.W.
- [1] On commuting rings of endomorphisms, Canad. J. Math. 8 (1956). 271-291.

DEMEYER, F.R.
- [1] On automorphisms of separable algebras II, Pacific J. Math. 32 (1970), 621-631.
- [2] Projective modules over central separable algebras, Canad. J. Math. 21 (1969). 39-43.
- [3] The Brauer groups of some separably closed rings, Osaka J. Math. 3 (1966), 201-204.
- [4] The Brauer group of a ring modulo an ideal, To appear.

DEMEYER, F.R. and INGRAHAM, E.
- [1] Separable Algebras Over Commutative Rings, Lecture Notes in Mathematics No. 181. Springer Verlag, Berlin, 1971.

DEURING, M.
 [1] Einbettung von Algebren in Algebren mit kleinerem
     Zentrum, J. Reine Angew. Math. 175 (1936). 124-128.
 [2] Algebren (2nd ed.), Ergebnisse der Mathematik und ihre
     Grenzgebiete, Springer Verlag, Berlin, 1968.

DIEUDONNÉ, J.
 [1] Compléments à trois articles antérieurs, Bull. Soc.
     Math. France 74 (1946), 59-68.
 [2] La théorie de Galois des anneaux simples et semi-
     simples, Comment. Math. Helv. 21 (1948), 154-184.

DOBBS. D.E.
 [1] Rings satisfying $x^{n(x)} = x$ have trivial Brauer group,
     Bull. London Math. Soc. 5 (1973), 176-178.

EICHLER, M.
 [1] Über die Idealklassenzahl total definiter
     Quaternionalgebren, Math. Z. 43 (1937), 102-109.
 [2] Über die Idealklassenzahl hyperkomplexer Systeme,
     Math. Z. 43 (1937), 481-494.
 [3] Bestimmung der Idealklassenzahl in gewissen normalen
     einfachen Algebren, J. Reine Angew. Math. 176 (1937),
     192-202.

EILENBERG, S. and MACLANE, S.
 [1] Cohomology and Galois theory, I. Normality of algebras
     and Teichmullers cocycle, Trans. Amer. Math. Soc.
     64 (1948), 1-20.

ENDO, S. and WATANABE, Y.
 [1] On separable algebras over a commutative ring, Osaka
     J. Math. 4 (1967), 233-242.

FADDEEV, D.K.
 [1] Simple algebras over a field of algebraic functions of
     one variable, Trudy Mat. Inst. Steklov 38 (1951),
     321-344; Amer. Math. Soc. Transl. Ser. II 3 (1956),
     15-38.

FEIN, B.
 [1] A note on the Brauer-Speiser theorem, Proc. Amer. Math.
     Soc. 25 (1970), 620-621.

FEIN, B. and SCHACHER, M.M.
 [1] Embedding finite groups in rational division algebras.
     I, J. Algebra 17 (1971), 412-428.
 [2] Embedding finite groups in rational division algebras.
     II, J. Algebra 19 (1971), 131-139.

FIELDS, K.L.
 [1] On the Brauer-Speiser Theorem, Bull. Amer. Math. Soc.
     77 (1971), 223.
 [2] On the Schur subgroup, Bull. Amer. Math. Soc. 77
     (1971), 477-478.

FIELDS, K.L. and HERSTEIN, I.N.
    [1] On the Schur subgroup of the Brauer group, J. Algebra, 20 (1972), 70-71.

FISCHER-PALMQUIST, J.
    [1] The Brauer group of a closed category, To appear Proc. Amer. Math. Soc.

FONTAINE, J.-M.
    [1] Sur la décomposition des algèbres de groupes, Ann. Sci. École Norm. Sup. (4) 4 (1971), 121-180.

FORD, C.
    [1] Pure, normal maximal subfields for division algebras in the Schur subgroup, Bull. Amer. Math. Soc. 78 (1972), 810-812.

FOSSUM, R.
    [1] The Noetherian different of projective orders, Thesis, Univ. of Mich. 1965.

FOSSUM, R., GRIFFITH, P. and REITEN, I.
    [1] The homological algebra of trivial extensions of abelian categories with application to ring theory, Preprint, Aarhus Univ., Mat. Inst., 1972/73.

FRÖHLICH, A. and WALL, C.T.C.
    [1] Equivariant K-Theory and algebraic number theory, Lecture Notes in Mathematics No. 108, Springer-Verlag, Berlin, 1969, 12-27.
    [2] Generalizations of the Brauer group, To appear.

GAMST, J. and HOECHSMANN, K.
    [1] Quaternions généralisés, C.R. Acad. Sci. Paris 269, (1969) 560-562.

GARFINKEL, G.S.
    [1] Amitsur cohomology and an exact sequence involving Pic and the Brauer group, Thesis, Cornell Univ. 1968.
    [2] A torsion version of the Chase-Rosenberg exact sequence, To appear.

GIRAUD, J.
    [1] Cohomologie Non-abélienne, Grundlehren der Mathematischen Wissenschaften in Einzeldarstellungen No. 179, Springer-Verlag, Berlin, 1971.

GLASSMIRE, W.
    [1] A note on noncrossed products, J. Algebra 31 (1974), 206-207.

GOLDMAN, O.
   [1] Quasi-equality in maximal orders. J. Math. Soc. Japan, 13 (1961), 371-376.

GREENBERG, M.J.
   [1] Lectures on Forms in Many Variables, Benjamin, New York, 1969.

GROTHENDIECK, A.
   [1] Le groupe de Brauer I, Algèbres d'Azumaya et interprétations diverses, Dix Exposés sur la Cohomologie des Schémas, North Holland, Amsterdam, 1968, 46-66 (Sém. Bourbaki 1964/65, exposé 290).
   [2] Le groupe de Brauer II, Théorie cohomologique, Dix Exposés sur la Cohomologie des Schémas, North Holland, Amsterdam, 1968, 67-87 (Sém. Bourbaki 1965/66, exposé 297).
   [3] Le groupe de Brauer III: Exemples et compléments, Dix Exposés sur la Cohomologie des Schémas, North Holland, Amsterdam, 1968, 88-188 (IHES March 1966).

HARADA, M.
   [1] Some criteria for hereditarity of crossed products, Osaka J. Math. 1 (1964), 69-80.

HARPER, L.R.
   [1] Differentiably simple algebras, Trans. Amer. Math. Soc. 100 (1961), 63-72.

HASSE, H. and SCHILLING, O.
   [1] Die Normen aus einer normalen Divisionsalgebra, J. Reine Angew. Math. 174 (1936), 248-252.

HATTORI, A.
   [1] Semi-simple algebras over a commutative ring, J. Math. Soc. Japan 15 (1963), 404-419.

HERSTEIN, I.N.
   [1] Noncommutative Rings, Carus Math. Monographs No. 15, Math. Assoc. of America, 1968.

HOCHSCHILD, G.
   [1] Automorphisms of simple algebras, Trans. Amer. Math. Soc. 69 (1950), 292-301.
   [2] Restricted Lie algebras and simple associative algebras of characteristic p, Trans. Amer. Math. Soc. 80 (1955), 135-147.
   [3] Simple algebras with purely inseparable splitting field of exponent one, Trans. Amer. Math. Soc. 79 (1955), 477-489.

HOECHSMANN, K.
   [1] Simple algebras and derivations, Trans. Amer. Math. Soc 108 (1963), 1-12.
   [2] Algebras split by a given purely inseparable field, Proc. Amer. Math. Soc. 14 (1963), 768-776.

HOOBLER, R.T.
- [1] Non-abelian sheaf cohomology by derived functors, Category Theory, Homology Theory and their Applications, Vol. III, Lecture Notes in Mathematics, No. 99, Springer-Verlag, Berlin, 1969.
- [2] Brauer groups of abelian schemes, Ann. Sci. École Norm. Sup. (4) 5 (1972), 45-70.
- [3] Cohomology in the finite topology and Brauer groups, Pacific J. Math. 42 (1972), 667-679.
- [4] A generalization of the Brauer group and Amitsur cohomology, Thesis, Univ. of Cal., Berkeley, 1966.
- [5] Purely inseparable Galois theory, Ring Theory - Proceedings of the Oklahoma Conference, Lecture Notes in Pure and Applied Mathematics, Marcel Dekker Inc., New York, 1974, 207-240.

HOOD, J.M.
- [1] Central simple p-algebras with purely inseparable subfields, J. Algebra 17 (1971), 299-301.

JACOBSON, N.
- [1] Abstract derivations and Lie algebras, Trans. Amer. Math. Soc. 42 (1937), 206-224.
- [2] p-algebras of exponent p, Bull. Amer. Math. Soc. 43 (1937), 667-670.
- [3] Lectures in Abstract Algebra, Vol. I, Van Nostrand, Princeton, 1951.
- [4] Structure of Rings, Amer. Math. Soc. Colloquium Publications, Vol. 37, 1956.

JANUSZ, G.J.
- [1] Separable algebras over commutative rings, Trans. Amer. Math. Soc. 122 (1966), 461-479.
- [2] The Schur index and roots of unity, Proc. Amer. Math. Soc. 35 (1972), 387-388.
- [3] Simple components of $\mathbb{Q}[SL(2,q)]$, To appear.
- [4] Generators for the Schur group of local and global fields, To appear.

KANZAKI, T.
- [1] On commutor rings and Galois theory of separable algebras, Osaka J. Math. 1 (1964), 103-115.
- [2] On Galois algebra over a commutative ring, Osaka J. Math. 2 (1965), 309-317.
- [3] On generalized crossed product and Brauer group, Osaka J. Math. 5 (1968), 175-188.
- [4] A note on abelian Galois algebra over a commutative ring, Osaka J. Math. 3 (1966), 1-6.

KAPLANSKY, I.
- [1] Fröhlich's local quadratic forms, J. Reine Angew. Math. 239/240 (1969), 74-77.
- [2] Commutative Rings, Allyn and Bacon, Boston, 1970.

KAROUBI, M.
   [1] Algèbres de Clifford et K-Théorie, Ann. Sci. École Norm. Sup. (4) 1 (1968), 161-270.

KATZ, V.J.
   [1] The Brauer group of a regular local ring, Thesis, Brandeis Univ., 1968

KNUS, M.-A.
   [1] Algebras graded by a group, Lecture Notes in Mathematics No. 92, Springer-Verlag, Berlin, 1969.
   [2] Algèbres d'Azumaya et modules projectifs, Comment. Math. Helv. 45 (1970), 372-383.
   [3] Sur le théoreme de Skolem-Noether et sur les dérivations des algèbres d'Azumaya, C.R. Acad. Sci Paris Sér. A 270 (1970), 637-639.

KNUS, M.-A. and OJANGUREN, M.
   [1] Sur le polynôme charactéristique et les automorphismes des algèbres d'Azumaya, Ann. Scuola Norm. Sup. Pisa 26 (1972), 225-231.
   [2] Sur quelques applications de la théorie de la descente a l'étude du groupe de Brauer, Comment. Math. Helv. 47 (1972), 532-542.
   [3] A note on the automorphisms of maximal orders, J. Algebra 22 (1972), 573-577.
   [4] A Mayer-Vietoris sequence for the Brauer group, To appear.
   [5] Théorie de la Déscente et Algèbres d'Azumaya, Lecture Notes in Mathematics No. 389, Springer-Verlag, Berlin, 1974.

KUYK, W.
   [1] Generic construction of non-cyclic division algebras, J. Pure Appl. Algebra 2 (1972), 121-130.
   [2] The construction of a large class of division algebras, To appear.

LANG, S.
   [1] Algebra, Addison-Wesley, Reading, Mass., 1965.

LONG, F.W.
   [1] The Brauer group of dimodule algebras, J. Algebra 30 (1974), 559-601.
   [2] A generalization of the Brauer group of graded algebras, To appear Proc. London Math. Soc.

MACLANE, S. and SCHILLING, O.F.G.
   [1] A formula for the direct product of crossed product algebras, Bull. Amer. Math. Soc. 48 (1942), 108-114.

MAASS, H.
   [1] Beweis des Normensatzes in einfachen hyperkomplexen Systemen, Abh. Math. Sem. Univ. Hamburg 12 (1937), 64-69.

MAGID, A.K.
   [1] Pierce's representation and separable algebras, Illinois J. Math. 15 (1971), 114-121.
   [2] Ultrafunctors, To appear.

MANDELBERG, K.I.
   [1] Amitsur cohomology for certain extensions of rings of algebraic integers, To appear.

MANIN, Y.I.
   [1] Le groupe de Brauer-Grothendieck en géométrie Diophantienne, Actes du Congrès International des Mathématiciens 1970, Tome 1, Gauthier-Villars, Paris, 1971.

MICALI, A. and VILLAMAYOR, O.F.
   [1] Sur les algèbres de Clifford, Ann. Sci. École Norm. Sup. (4) 1 (1968), 271-304.
   [2] Sur les algèbres de Clifford, II, J. Reine Angew. Math. 242 (1970), 61-90.
   [3] Algèbres de Clifford et groupe de Brauer, Ann. Sci. École Norm. Sup. (4) 4 (1971), 285-310.

MILNE, J.S.
   [1] The Brauer group of a rational surface, Invent. Math. 11 (1970), 304-307.

MILNOR, J.
   [1] Introduction to Algebraic K-Theory, Princeton University Press, Princeton, N.J., 1971.

MORRIS, R.A.
   [1] On the Brauer group of $\mathbb{Z}$, Pacific J. Math. 39 (1971) 619-630.

MUMFORD, D.
   [1] Introduction to Algebraic Geometry (preliminary version of first three chapters), Math. Dept., Harvard Univ., 1967.

NAGATA, M.
   [1] Local Rings, Interscience Tracts in Pure and Applied Mathematics No. 13, John Wiley & Sons, New York, 1962.

NAKAYAMA, T.
   [1] Divisionalgebren über diskret bewerteten perfekten Körpern, J. Reine Angew. Math. 178 (1937), 11-13.

NORTHCOTT, D.
   [1] A note on polynomial rings, J. London Math. Soc. 33 (1958), 36-39.
   [2] An Introduction to Homological Algebra, University Press, Cambridge, 1960.

OJANGUREN, M.
    [1] A non-trivial locally trivial algebra, J. Algebra 29 (1974), 510-512.

OJANGUREN, M. and SRIDHARAN, R.
    [1] Cancellation of Azumaya algebras, J. Algebra 18 (1971), 501-505.

OKADA, T. and SAITO, R.
    [1] A generalization of P. Roquette's theorems, Hokkaido Math. J. 1 (1972), 197-205.

PAREIGIS, B.
    [1] Über normale, zentrale, separable Algebren und Amitsur-Kohomologie, Math. Ann. 154 (1964), 330-340.

PERLIS, S.
    [1] Cyclicity of division algebras of prime degree, Proc. Amer. Math. Soc. 21 (1969), 409-411.

PICCO, D.J. and PLATZECK, M.I.
    [1] Graded algebras and Galois extensions, Bol. Un. Mat. Argentina 25 (1971), 401-415.

POSNER, E.C.
    [1] Prime rings satisfying a polynomial identity, Proc. Amer. Math. Soc. 11 (1960), 180-183.

PROCESI, C.
    [1] Rings with Polynomial Identities, Marcel Dekker Inc., New York, 1973.
    [2] On a theorem of M. Artin, To appear in J. Algebra.

RAMRAS, M.
    [1] Splitting fields and separability, Proc. Amer. Math. Soc. 38 (1973), 489-492.

RAYNAUD, M.
    [1] Anneaux Locaux Henséliens, Lecture Notes in Mathematics No. 169, Springer-Verlag, Berlin, 1970.

REINER, I.
    [1] Maximal Orders, mimeographed notes, Dept. of Mathematics, Univ. of Illinois, 1969.

RIM, D.
    [1] An exact sequence in Galois cohomology, Proc. Amer. Math. Soc. 16 (1965), 837-840.

ROQUETTE, P.
- [1] Splitting of algebras by function fields in one variable, Nagoya Math. J. 27 (1966), 625-642.
- [2] On Galois cohomology of the projective linear group and its application to the construction of generic splitting fields of algebras, Math. Ann. 150 (1963), 411-439.
- [3] Isomorphisms of generic splitting fields of simple algebras, J. Reine Angew. Math. 214/215 (1964), 207-226.

ROSENBERG, A. and ZELINSKY, D.
- [1] On Amitsur's complex, Trans. Amer. Math. Soc. 97 (1960), 327-356.
- [2] Automorphisms of separable algebras, Pacific J. Math. 11 (1961), 1109-1117.

ROSSET, S.
- [1] Locally inner automorphisms of algebras, J. Algebra 29 (1974), 88-97.

ROY, A. and SRIDHARAN, R.
- [1] Higher derivations and central simple algebras, Nagoya Math. J. 32 (1968), 21-30.
- [2] Derivations in Azumaya algebras, J. Math. Kyoto Univ. 7 (1967), 161-167.

SAMUEL, P.
- [1] Théorie Algébrique des Nombres, Hermann, Paris, 1967.

SCHACHER, M.M.
- [1] Subfields of division rings I, J. Algebra 9 (1968), 451-477.
- [2] Subfields of division rings II, J. Algebra 10 (1968), 240-245.
- [3] More on the Schur subgroup, Proc. Amer. Math. Soc. 31 (1972), 15-17.

SCHACHER, M.M. and SMALL, L.W.
- [1] Noncrossed products in characteristic p, J. Algebra 24 (1973), 100-103.

SCHARLAU, W.
- [1] Über die Brauer-Gruppe eines algebraischen Funktionenkörpers in einer Variablen, J. Reine Angew. Math. 239/240 (1969), 1-6.
- [2] Über die Brauer-Gruppe eines Hensel-Körpers, Abh. Math. Sem. Univ. Hamburg 33 (1969), 243-249.

SCHILLING, O.F.G.
- [1] Arithmetic in a special class of algebras, Ann. of Math (2) 38 (1937), 116-119.

SERRE, J.P.
- [1] Applications algébriques de la cohomologie des groupes. Théorie des algèbres simples, Sém. H. Cartan, Paris 1950/51, exposés 5-7.
- [2] Corps Locaux, Hermann, Paris, 1962.
- [3] Cohomologie Galoisienne, Lecture Notes in Mathematics No. 5. Springer-Verlag, Berlin, 1964.
- [4] Algèbre Locale: Multiplicités, Lecture Notes in Mathematics No. 11, Springer-Verlag, Berlin, 1965.

SILVER, L.
- [1] Tame orders, tame ramification and Galois cohomology, Illinois J. Math. 12 (1968), 7-34.

SMALL, C.
- [1] The Brauer-Wall group of a commutative ring, Trans. Amer. Math. Soc. 156 (1971), 455-491.

STOLBERG, H.J.
- [1] Azumaya algebras and derivations. To appear.

SUPRUNENKO, D. and TYSHKEVICH, R.
- [1] Commutative Matrices, Academic Press, New York, 1968.

SWEEDLER, M.E.
- [1] Multiplication alteration by 2-cocycles, Illinois J. Math. 15 (1971), 302-323.
- [2] Groups of simple algebras, To appear.
- [3] Amitsur cohomology and the Brauer group, International Congress of Mathematicians, 1974.

SZETO, G.
- [1] On the Wedderburn Theorem, Can. J. Math. 25 (1973), 525-530.
- [2] On a class of projective modules over central separable algebras, Canad. Math. Bull. 14 (1971), 415-417.
- [3] On a class of projective modules over central separable algebras II, Canad. Math. Bull. 15 (1972), 411-415.
- [4] The Azumaya algebra of some polynomial closed rings, Indiana Univ. Math. J. 22 (1973), 731-737.

TATE, J.
- [1] On the conjectures of Birch and Swinnerton-Dyer and a geometric analog, Dix Exposés sur la Cohomologie des Schémas, North Holland, Amsterdam, 1968, 189-214 (Sém. Bourbaki 1965/66, exposé 306).

TEICHMÜLLER, O.
- [1] Verschränkte Produkte mit Normalringen, Deutsche Math. 1 (1936), 92-102.
- [2] p-Algebren, Deutsche Math. 1 (1936), 362-388.
- [3] Über die sogenannte nichtkommutative Galoissche Theory und die Relation..., Deutsche Math. 5 (1940), 138-149.

TSEN, C.
    [1] Divisionalgebren über Funktionenkörpern, Nach. Ges. Wiss. Göttingen 1933, 335-339.

VAN OYSTAEYEN, F.M.J.
    [1] Pseudo-places of algebras and the symmetric part of the Brauer group, Thesis, Univ. of Amsterdam, 1972.
    [2] On pseudo-places of algebras, Bull. Soc. Math. Belg. 25 (1974), 139-159.
    [3] The p-component of the Brauer group of a field of characteristic p, Nederl. Akad. Wetensch. Proc. Ser. A, 77 (1974), 67-76.
    [4] Generic division algebras, To appear.

VILLAMAYOR, O. and ZELINSKY, D.
    [1] Galois theory for rings with finitely many idempotents, Nagoya Math. J. 27 (1966), 721-731.

WALL, C.T.C.
    [1] Graded Brauer groups, J. Reine Angew. Math. 213 (1963), 187-199.
    [2] Graded algebras, anti-involutions, simple groups and symmetric spaces, Bull. Amer. Math. Soc. 74 (1968), 198-202.

WATANABE, Y.
    [1] Simple algebras over a complete local ring, Osaka J. Math. 3 (1966), 13-20.

WATERHOUSE, W.C.
    [1] A non-criterion for central simple algebras, Illinois J. Math. 17 (1973), 73-74.

WILLIAMSON, S.
    [1] Crossed products and maximal orders, Nagoya Math. J. 25 (1965), 165-174.

WITT, E.
    [1] Über ein Gegenbeispiel zum Normensatz, Math. Z. 39 (1935), 462-467.
    [2] Zyklische Körper und Algebren der Charakteristik p vom Grad p^n, J. Reine Angew. Math. 176 (1936), 126-140.
    [3] Schiefkörper über diskret bewerteten Körpern, J. Reine Angew. Math. 176 (1936), 153-156.
    [4] Die algebraische Struktur des Grupenringes einer endlichen Gruppe über einen Zahlkörper, J. Reine Angew. Math. 190 (1952), 231-245.

WOLF, P.
    [1] Algebraische Theorie der Galoisschen Algebren, Math. Forschungsberichte, Vol. 3, Deutsche Verlag der Wissenschaften, Berlin, 1956.

YAMADA, T.
- [1] On the Schur index of a monomial representation, Proc. Japan Acad. 45 (1969), 522-525.
- [2] On the group algebras of metabelian groups over algebraic number fields I., Osaka J. Math. 6 (1969), 211-228.
- [3] On the group algebras of metabelian groups over algebraic number fields, II., J. Fac. Sci. Univ. Tokyo, Sect. I 16 (1969), 83-90.
- [4] Characterization of the simple components of the group algebras over the p-adic number field, J. Math. Soc. Japan 23 (1971), 295-310.
- [5] Central simple algebras over totally real fields which appear in $\mathbb{Q}[G]$, J. Algebra 23 (1972), 382-403.
- [6] The Schur subgroup of the Brauer group, Lecture Notes in Mathematics No. 397, Springer-Verlag, Berlin, 1974.

YAMAZAKI, K.
- [1] On projective representations and ring extensions of finite groups, J. Fac. Sci. Univ. Tokyo, Sect. I, 10 (1964), 147-195.

YOKOGAWA, K.
- [1] Brauer groups of algebraic function fields and their adèle rings, Osaka J. Math. 8 (1971), 271-280.

YUAN, S.
- [1] On the theory of p-algebras and the Amitsur cohomology groups for inseparable field extensions, J. Algebra 5 (1967), 280-304.
- [2] Differentiably simple rings of prime characteristic, Duke Math. J. 31 (1964), 623-630.
- [3] Central separable algebras with purely inseparable splitting rings of exponent one, Trans. Amer. Math. Soc. 153 (1971), 427-450.
- [4] On the Brauer group of local fields, Ann. of Math. 82 (1965), 434-444.
- [5] p-algebras of arbitrary exponents, Proc. Nat. Acad. Sci. U.S.A. 70 (1973), 533-534.
- [6] Brauer groups for inseparable fields, To appear.

ZARISKI, O. and SAMUEL, P.
- [1] Commutative Algebra, Vol. I, Van Nostrand, Princeton, 1958.
- [2] Commutative Algebra, Vol. II, Van Nostrand, Princeton, 1958.

NOTATION

$(\frac{a,b}{R})$ 21

$[a,b]$ 21

$[A]$ 20

$A \sim B$ 19

$B(R)$ 20

$_2B(R)$ 22

$\delta(A)$ 26

$\Delta(f,S,G)$ 82

$e(A)$ 34

$f \sim g$ 82

$J(A)$ 4

$J_R(A)$ 40

N_{red} 60

$O_\ell(E)$ 65

$O_r(E)$ 65

$tr_R(P)$ 1

INDEX

References are to page numbers.

A^e 11
A^o 11
Amitsur cohomology   150
automorphism, inner   86
axe   141

$B(S/R)$ 31
Bass' theorem   108
Brauer group   19

C_o 95
C_1 95
cancellation property   103 ff, 108
Cayley-Hamilton theorem   37
Cech cohomology   153
central separable   15
central simple   25
coboundary   33, 82, 150
coycle   33, 82, 150
complete   49, 52
completion   54
conductor   70
connected   57
crossed product   34, 82

derivation   42
    inner   42
dimension   59, 67
    Krull   71
division ring component   25
double centralizer theorem   18

étale   135, 148, 162
exact   114

faithfully flat   111
faithfully projective   1
flat   111
    faithfully   111

Galois extension   81
global dimension   59, 67, 70

H^1 82, 86, 87, 150
H^2 33, 34, 35, 82, 86, 92, 153
Hilbert's Theorem 90   85

idempotent   51, 57
    separability   12
index   26
inner automorphism   86
inner derivation   42

Jacobson radical   4

Krull dimension   71

lattice   62

maximal order   62
Morita theory   8, 20

Nakayama's lemma   4
Noether's theorem   85
norm, reduced   60

opposite algebra   11
order   61
    left   65
    maximal   62
    right   65

perfect 99, 100, 107
Pic 86, 108
projective 1
    basis 1
    faithfully 1
pythogorean 22

Quat (R) 22
Quaternion algebras 21, 44
Quaternions 21, 29, 69

R-lattice 62
R-order 61
radical 4
rank 5
reduced norm 60
reduced trace 59
reflexive 64, 131
regular 59, 67, 100, 101, 131

separability idempotent 12
separable 11
    central 15
Serre's problem 109
Seshadri's theorem 108
Skolem-Noether theorem 27, 87
splitting 31, 123, 124, 137, 156

tPic 154
trace 59, 83
    reduced 59
Tsen's theorem 95, 96

Wedderburn's theorem 25, 29

QA
251.3
O 78

JAN 23 1976